The Day That Lightning Chased the Housewife
... and Other Mysteries of Science

The Day That Lightning Chased the Housewife ... and Other Mysteries of Science

EDITED BY
JULIA LEIGH AND DAVID SAVOLD

Foreword by Alan P. Lightman

Illustrations by David Povilaitis

Published by arrangement with
The American Association for the
Advancement of Science

Harper & Row, Publishers, New York
Grand Rapids, Philadelphia, St. Louis, San Francisco
London, Singapore, Sydney, Tokyo, Toronto

First PERENNIAL LIBRARY edition published 1989.

LIBRARY OF CONGRESS CATALOG CARD NUMBER 89-45255

ISBN 0-06-097263-7

90 91 92 93 **FG** 10 9 8 7 6 5 4 3

Contents

Preface

These essays were first published in slightly different form by the American Association for the Advancement of Science. The editors wish to thank Jane Alexander, Mary Elisabeth Challinor, Lynn Crawford Cook, Bonnie Gordon, Charles Lean, Diana Morgan, Heléne Goldberg Ross, and Perry Turner for help at various stages with this material. Most of all, we are indebted to Allen Hammond, without whom this book would not be possible.

The essays collected here are not about the triumphs of scientific research, but instead focus on what science *doesn't* know. Despite the dizzy pace of scientific discovery in the twentieth century—from Einstein's theory of relativity to Watson and Crick's elucidation of DNA's structure—there still remain a considerable number of questions that scientists cannot yet answer. That is the theme of this book, for such questions lie at the very heart of scientific endeavor.

Why are men bigger than women? How does a monarch butterfly migrate 2,000 miles or more to a winter refuge it has never seen? Where did the Moon come from? Each question is an invitation to ponder a mystery in nature. "The most beautiful experience we have," Einstein said, "is the mysterious. It is the fundamental emotion that stands at the cradle of true art and true science."

Foreword

Letter from the Earth

The Omni sat upon his throne, staring at the columns of the planet Gamma. On this planet all mysteries had been solved.

By and by, a messenger came, with a little news. "Great One, we have found life elsewhere in the universe. Living creatures inhabit a planet called Earth, fifty light units distant, in the constellation Riga. Based on their broadcasts, the Earthlings seem a backward people."

"I sense that these creatures were noticed some time ago," said the Omni.

"That is correct, Great One," replied the messenger. "However, the Earthlings are so primitive and ridiculous, we did not think them worth your attention." Having delivered his message, the messenger ate one of his legs, in the custom.

The Omni thought and yawned and then he spoke. "Go to this planet. Study it and its peoples, and report to me. Then return to my chamber." The messenger bowed and went to Earth.

In time, he began writing his report, which he sent by telepathic mail:

This is a strange place. There is nothing resembling it at home, except in our nurseries. The Earthlings are unimaginably far below us in knowledge and refinement.

Let me begin with their education. The schooling is endless. After a basic school, there are advanced schools, then more advanced schools. A major subject in these schools is history. Earthlings believe they can predict the future by studying their past, so they spend a large part of their education trying to fathom the previous actions of their fellows. Earthlings also study language—not one but many. Each nation and tribe insists on using a different language for communication.

After an Earthling has some rudimentary grasp of one language, he has only begun. Some of the Earthlings are middle-aged by the time they leave school. Even then, they have learned little and keep asking questions. For example, Earthlings love games and sports, but they can rarely predict the outcome of a game. They study the sizes and habits of the players, but never the physics. Wherever you go, you hear people asking each other "Do you think the Yankees will win tomorrow?" or "How many points did Bird score last night?" I will give another example. What these Earthlings fathom least is their own behavior, which they learn nothing about in their schools. It obsesses them. They are constantly asking "Why am I not happy?" or "Who am I?" Throughout their lives, they exchange large sums of money to discuss this confusion with each other, often while stretched in a flattened position. This makes them feel better.

Much Earthling behavior derives from their limitation to only two genders, called male and female. (They have not yet discovered gender proliferation.) With so few choices, the courtship is frantic and often extreme. You cannot go anywhere without seeing one sex advertising itself to the other—in newspapers, in video broadcasts, on large signs at the sides of the roads, even in food stores. You cannot sell liquid nourishment or transport machines unless they excite sexual desire. Earthlings will give up their jobs over sex. They will change clothes over sex. Males fight to the death over a particular female. Furthermore, each of the two sexes is totally mysterious to the other. Earthlings seem to encourage this state of affairs. Males and females intentionally dress differently, conceal body parts from each other, and rarely talk truthfully to each other.

I will describe some of the social groups. There is a faction of people who devote their professions solely to studying certain questions. Such people are called scientists. Many scientists spend their whole lives struggling with a single question they consider profound. For example, a great deal of worry has gone into what causes aging, and how does living memory work, and how does a solar system form. It is as our planet was thirty thousand units ago. On the other hand, these scientists, incredibly, pay no attention at all to the fifth force, which regularly projects their socks and pencils into hyperspace. They search for their missing socks for a few days, then shrug their shoulders and act as if the phenomenon never occured.

I have actually observed scientists in their habitat. They have crude devices to help them in their researches, but they never know if the devices will work or if they will find anything new. They look through their so-called telescopes and microscopes and record what they see and look again and hope they will understand more than they did before. Some scientists try to guess the physical laws with mathematics. They make theories. Then they revise their theories. Then they revise their revisions. In sum, they never know when they are finished. For example, these scientists have the notion that all matter and force derives from a small group of elementary particles. Yet they are constantly changing their minds about how many such particles exist. At one time there were 4 elementary particles, then there were 113, then 25, then 72. Sometimes, the scientists think that they are simplifying their systems with fewer elementary particles, sometimes with more. Not to mention their unquestioned assumption that the simplest theory is the one most likely true. And they never give up. They go to their rooms or their laboratories with their crude devices and they spend year after year in search of an answer. Once in a while, they make some small discovery and become excited, like children. Furthermore, for every question they answer, they uncover two more in the process.

Without doubt, Great One, the Earthling culture stems from the same ignorance and uncertainty. For example, a great many of the creatures never outgrow the childish habit of placing pigment on paper. Far into adulthood, they make images of what mystifies them or amazes them or agitates them, and a random few of these images are placed on the walls of museums and celebrated. Can you imagine paying money to see the scratchings of our children, before the middle datajump? In effect, that is what Earthlings do. They express their childlike mystery of the world in something they call art, which they have raised to the peak of their culture.

As with the scientists, the artists are never satisfied with their work and never know when it is finished. "That blue is not right," or "that green is too pale," or "that line should be darker." The painters keep dabbing on paint until they get tired, or the painting is bought, or it is crated up and moved out for more space in the studio. There is one painting I saw that serves as a perfect example. The artist couldn't decide whether to make the subject a male or a female, happy or sad, nearby or far. He placed the corners of the mouth in shadow, and because Earthlings are accustomed to not seeing anything finished, they consider this one of the greatest of all paintings.

There are other arts as well. From sound waves like our first synthesized polychords, the Earthlings have fashioned an art they call music. Musicians hear puzzling sounds in their heads, force these sounds into language, and then find the language inadequate. The most famous musician composed nine compound polychords, the last of which mixes Earthling voices with ancient music instruments. The composer was so dissatisfied with this mix that he decided to remove it, but died before he could. Yet the Earthlings consider this particular composition the best of its kind. Musicians repeatedly perform their music in concerts, before many people. Yet they never know exactly how their fingers will move or how the music will sound on a given performance. This uncertainty seems to please them and their audiences.

Some Earthlings vent their confusion by writing hundreds of pages on "love" and "sadness" and "greed"—so many names have they made for the chemical response of the root tropism. And these writers are called artists too. Their writings are read by other Earthlings and pondered for meaning. Like the painters, the writers are never satisfied with what they have written. They forever keep adding and subtracting words in their writings, until the manuscripts are physically wrenched away by the publishers. Afterwards, the writers are gloomy and complain that their work is unfinished. Often, years later, they go back and drastically revise their writings, having changed their minds about what they intended.

In some ways, the artists are stranger than the scientists. They are a boisterous bunch. They laugh and cry and publicly agonize over their ignorance. The rest of the population enjoys this display. I speak the truth, Great One.

Whatever artists don't understand, they turn into art. I have seen artists mindlessly gaze at the Earth's local star as it touches the horizon and then, with no grasp of what they have seen, make paintings and writings and ritualize their incomprehension. Some artists simply gaze at themselves and turn *that* into art.

These are my principal observations. I could say more, but I have seen enough of the Earthlings. Their restlessness tires me.

The messenger was finished with his report and returned to his planet. When he arrived, the Omni was sleeping, as were most other Gammans, stored in their sleeping compartments. The messenger walked through the silence, into the capitol.

"Great One, I regret to disturb you," whispered the messenger.

The Omni lazily opened his eyes and belched, then settled himself on his cushions. "You are back."

"I am back," said the messenger.

The Omni held in one palm a small glass ball, which he passed from one hand to the next, then back again, from one hand to the next, then back again. "You have returned."

"Correct."

"Your report is finished."

"Yes."

"But you have not described the children of Earth," said the Omni. "Tell me about the children, before I begin my nap."

"They are baffled by the world, like their parents," said the messenger. "Clearly, their questions are never answered. The adults give answers, but the children know the answers are unfinished. They learn to love the questions. And at night, they have odd visions in their sleep. They imagine horses with wings, underwater cities, giant mirrors, half opened doors."

The Omni nodded, closed his eyes, and returned to a dreamless sleep.

—*Alan P. Lightman*
Harvard-Smithsonian Center for Astrophysics

Part I
THE MIND

Why Do We Dream?

Each night we lie insensible to the world, our eyeballs bobbing beneath closed lids as scenes and stories flit across our brains. But what function does our dreaming serve? The question has nagged at humans for millennia.

In 1900, Freud wrote in *The Interpretation of Dreams* that science had contributed "little or nothing that touches upon the essential nature of dreams or that offers a final solution of any of their enigmas." By the early 1950s sleep researchers had found evidence for regular dream-packed periods of sleep, called REM sleep for the rapid eye movements that occur for a total of about two hours a night. Since then, the data have piled up. Yet Freud's turn-of-the-century observation could as well apply today.

Perhaps REM sleep clears the nervous system of metabolites deposited through the day. Others have suggested that dreaming is mainly useful in early development, perhaps helping to establish the brain circuitry the fetus and infant need to process information. Human newborns, it is known, spend about half their sleep time dreaming, compared to just 15 percent in 60-year-olds.

There is, in any case, no want of theories. Ubiquitous,

bizarre, offering rich grist for daybreak whisperings and the intimate revelations of the psychiatric couch, dreams fairly invite theories. Why, as sleep lab evidence shows, do we remember some dreams, forget others? Why their broken logic? Why such recurring themes as falling, or being chased? But most of all, why do we dream at all? An overarching synthesis remains elusive, frustrated not so much by conflicting evidence as by theoretical perspectives that scarcely bear on one another.

By one view, dreaming is simply a side effect of natural physiological processes. Every 90 minutes or so the pons, a concentration of nerves in the brain stem, issues signals to higher brain centers. These signals may help form the visual images and cause the characteristic rapid eye movements of dream sleep. By this reasoning, the bizarre imagery of dreams, the time compression and abrupt scene shifts, are just the brain doing what it can with the mishmash of signals originating in the pons.

But this neurophysiological theory doesn't explain the content of dreams. Psychologists see dreaming as a way of dealing with emotionally charged material that waking consciousness lacks time to deal with. Studies show, for example, that people in the midst of a stress-filled divorce have long, complex dreams, while others with placid lives report dull dreams. Freud had seen in dreams the expression of unconscious wishes, the "protection" of sleep from ugly or painful inner truths. And Jung viewed them as a window into the collective unconscious, rife with myth and symbol. Today the computer suggests entirely new perspectives. Can the older notions based on content survive insights from the recent spate of information processing models?

In 1967, Edmond Dewan first likened dreaming to "off-line processing"—as if a computer were reprogramming itself, without intervention from humans. The idea is that simple organisms may respond to their world in instinctive, or preprogrammed, ways. But mammals—all of which are thought to dream save an Australian anteater—first have to learn about their environment. Then, in dreams, with the outside world shut off, they may integrate their new knowledge, reprogramming and rehearsing their behavior in response.

If there's consensus among researchers, it's that dreaming must somehow serve learning. Dreams record experience, or assimilate emotional material, or consolidate memory—all variants on the theme.

Nobelist Francis Crick and a colleague expand learning to embrace "reverse learning." A neural network in the brain that stores information can theoretically become overloaded. Perhaps, they say, dreaming purges these networks of useless, trivial, or redundant patterns.

Just how dreaming might serve learning physiologically is not so clear as learning-based theories suggest. Jonathan Winson of Rockefeller University envisions the part of the brain known as the hippocampus as a gatekeeper for information, switching signals one way during wakefulness and dream sleep—both periods of high mental arousal—and another way the rest of the time. Is this the physiological equivalent of "off-line processing"? Is it, as he suggests, the neural basis for the Freudian unconscious?

Whatever its fate at the hands of experiment, Winson's theory at least seeks reconciliation—between Freud, biology, information processing, and the common sense that says dreams *mean.*

—*Robert Kanigel*

Why Do We Sleep?

"If sleep does not serve an absolutely vital function," says Allen Rechtschaffen in his book, *The Control of Sleep,*" then it is the biggest mistake the evolutionary process ever made."

The impulse to sleep is overwhelming: we spend a third of our lives doing it and feel terrible without enough. Yet after more than three decades of research, we still are not sure of its function. Ask a number of sleep researchers the reason we sleep, and the answer you are likely to hear is, "because we get tired."

Attempts to determine why we sleep have led researchers in many different directions. At Emory University in Atlanta, Gerald Vogel has conducted studies to determine the role of rapid eye movement (REM), or dreaming, sleep in our waking behavior. One early speculation was that if REM sleep were suppressed, it would lead inevitably to psychosis. The idea, long since abandoned, was that if you suppressed REM sleep, dreams would force themselves into waking behavior and result in a kind of "dream state" similar to some forms of schizophrenia.

Vogel tried the reverse: He treated a number of clinically

depressed patients by depriving them of their REM sleep for several weeks to see if their condition improved. No other drugs or treatments were used, and more than half of his patients showed significant and permanent recovery. Curiously, normal people in the study who were similarly deprived of REM sleep showed nothing more than a "REM rebound" effect signified by vivid dreaming once the REM deprivation ended.

What this suggests to Vogel is that there is some mechanism related to REM sleep that is linked to our "conscious enthusiasm." Depression may occur when something within this mechanism goes wrong, but precisely what or why remains a mystery, as does the apparent need for sleep in the first place.

Is sleep restorative? It certainly feels as if it is, but so far experiments designed to prove it have failed. Recently, researchers tried to determine if tissue restoration occurred during sleep. If so, protein, a major component of tissue, would be synthesized faster than it broke down during sleep. Precisely the opposite was found.

If restoration could not be proved directly, how about indirectly? Would the absence of sleep have serious and perhaps irreversible physiological consequences? All studies to date, which include keeping people awake for up to 12 days, indicate that nothing particularly harmful results. After a couple of good nights' sleep, these people feel like their old selves. In some cases, people without sleep have reported psychotic episodes, but sleep researchers believe these are caused by underlying psychological problems and not sleep deprivation itself.

Investigators have also tried to assess the role of the stages that measure sleep's depth. We know that as we grow older we sleep less deeply until we no longer reach the deepest stage at all. We also know that the pituitary gland releases growth hormone during the deepest stage. Growth hormone plays a vital role for children, but it may also be important for adults. Beginning in our mid-40s, as we sleep less and less in the deepest stage, the growth hormone release slows to a near halt. Sleep and its connection with growth hormone may be part of the larger bio-chemical process of aging but precisely how it interconnects is again unknown.

Two other hormones, prolactin and lutenizing hormone (LH), are also secreted during sleep. Prolactin plays an important role for nursing mothers, and LH is necessary for the production of sex hormones. But prolactin, LH, and growth hormone are secreted during wakefulness as well, though in much lesser amounts. Some as yet unknown restorative or hormonal process may occur during sleep, but it is certainly not unique.

What investigators continue to be left with are tantalizing clues, fragments of ideas that refuse to solidify into a coherent theory of why we sleep. A major obstacle is that we have not yet devised techniques sophisticated enough to probe deeply and subtly into the workings of the sleeping brain.

Many sleep researchers believe we will one day have those techniques. At present, the reason we sleep is elusive. But most researchers are betting against finding that sleep is "the biggest mistake the evolutionary process ever made."

—*John Pekkanen*

How Did Language Begin?

Cavemen would come to need what all languages provide—ways to distinguish between cause and effect, present and past, real and unreal, state and process. They'd need a way to show specificity, so that "Let's have lamb for supper" could not be confused with "Let's eat Fluffy." We can figure what cavemen would eventually need to say; anthropologists and physiologists who study ancient skulls can even guess when they came to say it. But there are no fossil clues to how language began.

Perhaps something could be deduced from animal communication. Chimpanzees, for example, have been taught to converse in sign language, and some theorists suggest that human language began as a system of gestures or with calls of alarm. Others object, as no one can conclusively trace a line from animal-like communications to the subtle, complex language humans use.

The majority of linguists—though by no means a consensus—believe humans' ability to learn language is at least in part innate, that it is somehow woven into our genetic makeup. Noam Chomsky, theorist at the Massachusetts Institute of Technology, proposes a specific neurological program that enables children to learn to use grammatically correct structures instead of nonsensical ones. Children, says Chomsky, cannot learn language solely on the basis of what they are taught, and in any case they learn it too quickly for that to be true.

But evidence for a genetic program is hard to come by. Children learn language in an intensely verbal environment, and linguists cannot maroon babies in a desert laboratory. Oddly, a few circumstances do resemble marooning. Some children born deaf will invent a system of signs their parents do not use. And more provocatively, children will use creoles, a class of languages that arise when speakers of mutually unintelligible languages are thrown into persistent contact.

Derek Bickerton, linguist at the University of Hawaii in Honolulu, believes that creoles provide evidence for an innate language program. Creoles—more than a hundred are known—generally appeared when the slave trade and European colonialism forced great numbers of people who spoke different languages to work together. Because no one, for example, spoke Haitian Creole French before 1630, nor Hawaiian Creole English before 1880, linguists have been able to look at not only the grammatical structure of creoles but also their development.

Typically, a community of workers would first speak a pidgin. They would use the vocabulary of their bosses with part of whatever grammar was native to them. Pidgins are makeshift, inefficient languages with grammatical features not common to all speakers. The workers' children, however, spoke a creole. The children's vocabulary didn't differ from pidgin, but they used a common grammar, which, according to Bickerton, is unlike any they could have heard. Remarkably, all creoles share similar grammar.

Creoles, then, appear to solve some of the same problems—tense, for example—in the same way. In Hawaiian

Creole English, "walk" is past tense only; "bin walk" means "had walked"; "stay walk;" is continuing action, as in, "I am walking" or "I was walking"; and "I bin stay walk" means "I had been walking." Very similar tense systems are used in all other creoles, whatever their vocabularies.

Languages from different times, in different countries, with different vocabularies that still have common grammatical features? How does that bear on the origin of language? According to some linguists it doesn't. Languages spread, of course, and creoles may have linguistic roots in the pidgins that arose as early Portuguese traders traveled through West Africa. Bickerton, however, sees creoles as linguistic newborns, and he believes that creole grammar reflects Chomsky's innate program for ordering language.

As Bickerton explains it, all children grow up in an environment relentlessly speaking Japanese, French, English, etc. But they draw on the same innate grammar that gives rise to creoles. Linguists have long known that as children learn their parents' language they make certain mistakes systematically. Dan Slobin of the University of California at Berkeley has studied children's mistakes across at least a dozen different languages. When comparing their errors to creole grammar, says Slobin, "the resemblance is striking." For example, many children, like creole speakers, use a negative subject with a negative verb: "Nobody don't like me," a child will say; or in Guyanese Creole, "No dog did not bite no cat." And neither child nor creole user asks questions by changing word order; they question by intonation alone: "You can fix this?"

In Bickerton's scenario, then, a caveman's language would have structurally resembled a creole. Succeeding generations of cavemen, like the developing child, would have suited language to environment. Different cultures make different demands on language. "Biological language," says Bickerton, "remained right where it was, while cultural language rode off in all directions." But when culture is uprooted—the slave trade, for example—our biological rules for ordering language are still at home.

Reactions to Bickerton's hypothesis are mixed. Not only might creoles be descended from other languages, but some linguists doubt that the child who uses a creole qualifies as "marooned," for he still grows up hearing language. And many people question Bickerton's interpretation of those grammatical similarities.

A few linguists do believe Bickerton is on to something. "On the whole," says John Rickford of Stanford University, whose native language is Guyanese Creole, "Bickerton might have his left sock on his right foot, but in the long run, I think he's going to be dressed." Even if there is an innate component to language, which isn't a new idea, it doesn't solve the mystery of how such a Chomsky-type program arose. The origins of language remain obscured in our evolutionary past.

—*Ann Finkbeiner*

Why Do We Like Music?

It started simply enough: just a pulse in the lowest registers—
bassoons and basset horns—like a rusty squeezebox. . . .
And then suddenly, high above it, sounded a single note on
the oboe. It hung there unwavering, piercing me through,
till breath could hold it no longer, and a clarinet withdrew
it out of me, and sweetened it to a phrase of such delight it
had me trembling.

Peter Shaffer, Amadeus

Music, after all, is nothing more than a sequence of sound
waves. So why did the music of Wolfgang Amadeus Mozart
fill his rival, Antonio Salieri, with the longing and pain
described in Peter Shaffer's play? Why do we fill our own
lives with music—in the concert halls of Vienna, in the
streets of Harlem, on the plains of India? What is it that
allows a sequence of sound waves to touch us so deeply?

Part of the answer seems to lie in the physics of the
sound waves. Scales and chords, for example, are con-
structed from pitches that are mathematical progressions of
one another. When we hear a middle C, the air is vibrating
some 260 times per second. Double that to 520 vibrations

per second, and we hear a C exactly one octave higher; multiply middle C's vibrations by 3/2, and we hear the G in that octave at 390 vibrations per second.

Over the centuries musicians have elaborated such relationships into an enormous body of music theory. But valuable as it is, theory only tells us how music works, not why. It cannot explain why one tune is utterly banal and another is magic.

Obviously, a great deal of our appreciation for music is learned. You may like a song that I hate simply because it resembles other songs that you like. On a more fundamental level, the aesthetics of music vary widely between cultures. In the Orient the stress is on pitch and tiny intricate intervals. In sub-Saharan Africa the rhythms reach dizzying complexity. In the 18th-century Europe of Bach and Mozart, the ideal was order, structure, and balance.

But again, none of this explains why almost everyone responds to some kind of music, or why music in one form or another appears in every known human society.

In the last analysis, it seems that the power of music lies not in the sounds but in ourselves. Just as our eyes are receptors to light and our ears are receptors to sound, we somehow have in our brains a receptor to music. In fact Harvard psychologist Howard Gardner argues that musical intelligence is something that is separate and coequal with other forms of intelligence, such as an ability with words or with numbers.

In many ways, Gardner says, music and language abilities are very similar. Babies start to babble fragments of "song" as early as they start to make little word sounds. Older children progress in stages, showing an ability to sing longer and more complex songs in much the same way as they start to use longer and more complex sentences.

But music is not just language in another form, says Gardner. For example, the Soviet composer V. Shebalin suffered a stroke in the left temporal lobe of his brain, the area for language comprehension. Afterwards he had great difficulty communicating, yet his compositions were as brilliant and as sensitive as ever.

On the other hand, a young musical composer suffered damage to the right hemisphere of his brain. He had no trouble communicating and eventually returned to teaching music. But he had lost all interest in composition. He even lost much of his enjoyment in listening to music.

Studies such as these, Gardner says, indicate that some key essence of our musicality is located in the right front of the brain. The exact location, however, and the exact nature of that essence is far from clear. Even if we do someday track down the brain's "music receptor," we are still left with one final mystery: Why is it there? Some scholars have suggested that our musical abilities evolved at the same time we acquired language, anywhere from a few hundred thousand years ago to a million years ago. Yet language gave our tribal ancestors a clear evolutionary advantage: Better communication meant a better chance at survival. What need did music serve?

Of course, we could also ask that question about painting or sculpture, dance or poetry. Why do humans respond to beauty of any kind? To that question, we have no more answer than Shaffer's tortured Salieri, who cried up to his 'sharp old God': "What is this? Tell me, Signore! What is this pain? What is this need in the sound? Forever unfulfillable, yet fulfilling him who hears it, utterly."

—*M. Mitchell Waldrop*

How Do We Remember?

It's perhaps the most tantalizing mystery confronting scientists whose research province is the brain. How does the brain learn and remember what is learned? It doesn't really help to know that the brain is "hard wired"—that, even before birth, its millions upon millions of individual nerve cells have already linked up according to a genetically predetermined and astonishingly complex master plan. How does experience impress itself upon such a rigid neural network?

We are not born knowing German or French or English or Urdu. We must learn language. Nor does a child know automatically to avoid the flame. He must learn—often painfully.

But how are these lessons planted in the brain circuitry? With the connections among neurons or nerve cells presumably invariant, and with no new connections being made, what is modified to accommodate new experience?

According to one hypothesis, memory was stored in "reverberating loops," based on an observation that neurons in the brain were often interconnected in the form of closed loops. In this scheme memory and learning were sustained by the circulation or reverberation of electrical signals within the loops; these round-and-round signals were the files, the data bank of memory.

Then there was the idea of "memory molecules." This notion, which borrowed from discoveries related to the breaking of the genetic code, captured the popular fancy for a while. It held that memories were stored in large molecules, with the bits of information making up the memory encoded in a sequence of smaller molecules— much as genetic information is encoded in DNA, the master chemical of heredity.

The theory now in vogue—the so-called plastic theory— focuses on the synapse, the junction or gap across which a nerve cell transmits its electrical impulses, via chemical intermediaries, or neurotransmitters, to a second nerve cell, or muscle. Thus learning and memory impress themselves on the nervous system by "strengthening" the synaptic linkage.

But how? And which pathways does the brain choose to strengthen? Does it select certain routes at the expense of others? And, for that matter, what is it that our brains choose to learn and remember?

Behavioral scientists have identified three distinct orders of memory. First, there's iconic memory, lasting no more than a tiny fraction of a second. Iconic memory is "photographic"—total, active recall of a perceived scene that's more than merely an afterimage on the retina. Then there's short-term memory, which persists for several seconds, when one looks up a phone number, for example, and remembers it just long enough to dial. Short-term memory is what William James once referred to as "immediate awareness" and what Richard Thompson, professor of psychobiology at the University of Southern California, describes as "what you have in your mind at the moment." Lastly, there is long-term memory.

But within the framework of this apparent learning hierarchy, what is it that the brain chooses to store permanently? Or does the brain, as some argue, actually hang on to everything that happens to it in life—much of it in the form of unrecallable memory? It's possible; there's no shortage of data storage capacity. According to Thompson, the human brain, with 50 billion nerve cells, has more possible synaptic connections than the total number of particles in the universe.

Others contend that the brain keeps only those things that it deems important, which presupposes the existence of a neural "gatekeeper" or "significance detector." But, then, what is it? Where is it? And how does it work?

—Ben Patrusky

Part II
BIOLOGY

How Do Cells Know What To Become?

Omnes ab ovo. Everything comes from the egg.

Uttered nearly three and a half centuries ago, William Harvey's dictum still remains at the root of the central question in biology: How does a single cell, the fertilized egg, give rise to a complex organism composed of billions of cells with very different specialized functions?

How is the original egg cell transformed into heart, blood, liver, brain, skin, kidney, and gut—into cells that differ in structure, in the proteins they make, and in the way they behave?

Answers are being sought at three different levels of cell organization: the nucleus, which houses the DNA, the stuff of genes; the cytoplasm, the jelly-like material surrounding the nucleus; and the membrane, the social director of the cell, which mediates its interaction with all the other cells in the environment.

The prime mover of differentiation, it would seem, is the DNA, the master chemical of heredity, carrier of the blueprint of cell development. What perplexes, however, is that the genetic information in the fertilized egg—the entire complement of DNA—gets copied and passed on to the daughter cells and all subsequent cell generations. But if every cell derived from the original egg bears the same genetic repertoire, how is it that cell lines can evolve so differently?

Currently, the only reasonable explanation for the specialization of cells has to do with the selective inactivation of certain groups of genes and the activation of others. In other words, different cell lines for heart, brain, or skin, for example, make use of different parts of the repertoire. But then what is the nature of the mechanism that causes various genes to express themselves and others to keep mum? One promising lead stems from the recent discovery of "jumping genes"—movable fragments of DNA—which seemingly have the wherewithal to serve as the genetic "on-off" switches.

But such speculations, even if borne out, only breed other riddles: What decides which sets of genes should be turned on and off when? After all, embryos develop according to a highly formalized, rigorously timed schedule; differentiation—cell change—must proceed in an exquisitely precise, clocklike fashion. Thus there must be some sort of biological foreman seeing to it that genes become activated or get suppressed at exactly the right time and in the proper sequence.

The master control, curiously, may actually lie outside of the nucleus—in the cytoplasm. That may come as a surprise inasmuch as genes govern the manufacture of substances that make up the cell cytoplasm. Nevertheless, and never mind the paradox, the converse may also be true: Not only does the nucleus control the cytoplasm, but the cytoplasm may also regulate the nucleus.

Supportive evidence for this idea comes from studies showing that cytoplasm is not homogeneous, not even in

the original egg cell. There are, instead, regional differences: Cytoplasm in one part of the cell differs from that in
another. When the egg cell cleaves, the cytoplasmic distribution is not identical; each daughter cell receives a different allotment of cytoplasmic goods. Thus these cytoplasmic
differences, by affecting the nucleus in different ways, may
serve as the earliest instruments of differentiation.

And then there is the cell membrane. It, too, has a crucial
role in the differentiation process. The membrane gives the
cell social identity, enabling it to recognize and interact
with other cells in its environment. The cell's social charac-
ter derives from surface chemicals, particularly surface
proteins. As cells diverge, they develop distinctly different
surface-protein "signatures." These signatures are spelled
out by the cell nucleus, which directs the manufacture of
protein kind, and the cytoplasm, which plays a major part
in properly positioning the proteins on the membrane.

But it's a two-way street: The membrane, in turn, has a
telling effect on cell function. It receives signals from other
cells and from the extracellular environment—and these
signals are subsequently transmitted to the cytoplasm and
nucleus so as to modify cell development and behavior.

Obviously, from what's now known, it's impossible to
label any one of the cellular subunits—the nucleus, the
cytoplasm, or the membrane—as the key element in differentiation. Each is essential; each strongly affects the others,
but in ways that will likely remain deep, dark secrets for a
long time to come.

Omnes ab ovo. Everything comes from the egg, all
right—including one heck of a biological mystery.

—Ben Patrusky

The How And Y Of Maleness

There are disagreements over the precise ingredients, but everyone knows that to make both boys and men you start with a Y chromosome. At the moment of conception, Y meets X, and together they become one of 23 pairs of chromosomes carried in the cells of human males. What remains puzzling is just how the Y, a tiny string of genes compared to its partner, triggers the development of the male sex.

Early in our evolutionary history, the Y chromosome, for reasons not known, passed most of its genes to the X and shrank in the process. Though the male chromosome ended up standing in the shadow of its partner, it held on to one critical power: the determination of maleness.

Gender doesn't begin to emerge in the embryo until the second month of development. Then, in response to some mysterious signal, cells form that will become ovaries or testes. If the egg, which always bears a single X chromosome, has been fertilized by an X-bearing sperm, a female—an XX—will develop. But if the sperm carried a Y chromosome, the baby will be male, an XY.

Unlike larger chromosomes, whose genes have been

mapped, the geography of the Y remains largely unexplored. Still, somewhere along the Y's two small arms is the male-determining portion. "We suspect it's either a gene or a cluster of genes in physical proximity," says David C. Page, a geneticist at the Whitehead Institute in Cambridge, Massachusetts. But as yet, no one knows the trigger's precise location. Nor do they know exactly how the trigger operates.

Page and colleagues in France and Finland are looking for the trigger in a seemingly strange place: along certain X chromosomes. About one of every 20,000 men, it turns out, has two X chromosomes instead of an X and a Y. These XX men are infertile, but often they don't know it until they try to become fathers and fail.

"Each XX male is the result of a genetic lightning strike," says Page. The chromosomes of the father and mother may be perfectly normal, but when sperm are created an accident occurs. Normally, during a process called meiosis, all 23 pairs of chromosomes separate, so that some sperm end up carrying a Y, while others carry an X. During this division, chromosomes often swap bits of information. It's rare for the Y to participate in such reshuffling, but it seems to have occurred in XX males.

Using a sophisticated probe, Page and colleagues have found telltale Y-derived DNA sequences incorporated into what they suspect is the paternal X chromosome of XX men. Just what these gene sequences code for is unknown, but they may well contain the key information for making a man. It could be directions for a hormone, a protein, or an enzyme.

One substance under suspicion is a blood factor known as the H-Y antigen. Discovered in 1955, H-Y has since been found in many mammals—mostly, but not exclusively, in males.

"The H-Y antigen may be the long-sought determinant of testicular differentiation," says immunogeneticist Stephen

S. Wachtel, director of the Reproductive Immunogenetics Laboratory at the University of Tennessee, Memphis. He explains that the antigen causes "indifferent" tissues—tissues that can become either male or female sex organs—to differentiate into precursor cells that will ultimately make the structures and hormones that make the man. Without this antigen, the cells follow a female course.

The system, however, is not foolproof. Sometimes the precursor cells can't respond to the H-Y antigen, so they wind up developing into abnormal ovaries. An XY female, like an XX male, is infertile.

Not everyone agrees that the H-Y gene is the trigger for maleness. In one study, XX male mice were found to lack the H-Y antigen. Even more confusing, there maybe more than one H-Y antigen, and there is now good reason to believe that the H-Y gene isn't even on the Y chromosome. Instead, it may be somewhere on the X. If the antigen turns out to be the key substance, the Y's role in determining masculinity may be relegated to that of a bit player—an on-off button that somehow merely activates the critical gene or genes.

"We just don't know how the Y flips the switch," says Page. Until then, men remain mysteries.

—*Bruce Fellman*

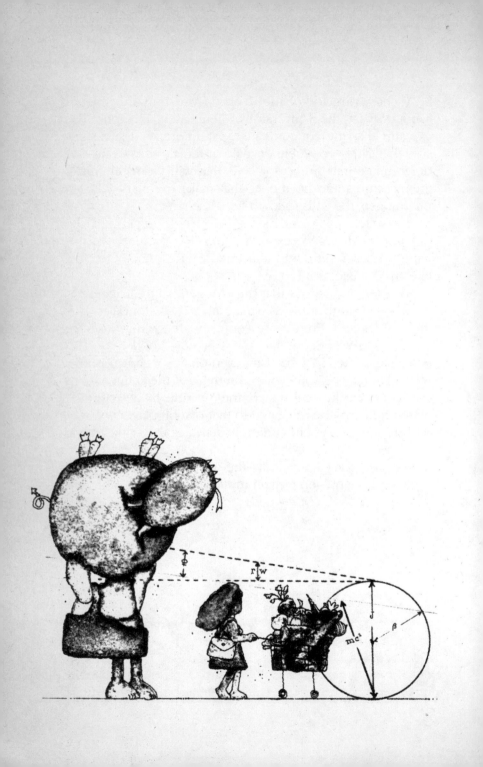

Why Are Men Bigger Than Women?

According to many creation myths that account for the origin of our species, the size difference between the sexes is all part of the plan. In the beginning, men were made the bigger animal because the creator cast them in the role of provider and protector.

But the matter is not so easily resolved by scientists. Some have suggested that the larger male was more successful in competition with other males for females. But other investigations point out that the female, then, also chose that bigger mate. Anthropologists are divided—often bitterly—on this "who chose whom, and why" point. Though no one knows its origins, size inequality must have served a purpose and been worthy of perpetuation.

Not all theories of why men are bigger than women focus on choice. One idea is that the less energy a female needs to expend, the better her chances of perpetuating the species. A smaller female carrying and nursing her offspring would require as many calories as her mate. If you assume that food was generally in short supply through most of human history, she couldn't have been as big as her mate.

So how did our ancestors go about conserving their reproductive energies? Perhaps, several million years ago, the male did most of the foraging to provide for the family. The more stationary the female, the less chance of losing her children to predators. And if, suggests Owen Lovejoy at Kent State University, the male primarily mated with only one female, he might reasonably be assured that the food he brought back went to offspring carrying his genes. Strength and size, then, helped the more mobile male fend off predators, while the female's smaller size helped her hide from them.

Another theory is that our ancestors practiced "niche differentiation"; that is, they divided up the labor required to feed themselves along male and female lines. Some 500,000 to 700,000 years ago, men probably hunted while women gathered. By that time there were tools to help deal with big game, and hunting might have provided a reliable food source. Large size could have contributed to a hunter's success and advertised his prowess.

By the time such a role separation existed, however, the size difference, or dimorphism, between the sexes had shrunk to the level it is today, about 10 to 20 percent. And it is argued, judging from other primates, that there is no reason to believe that early hominids were monogamous. So what purpose did the size difference serve? Why are men still bigger than women?

A look at primates suggests that breeding strategies may provide clues to the mystery. The male gorilla, for example, is about twice the size of the female. He competes with other males for control of a group of foraging females—a harem strategy in which male size can determine competitive and reproductive success. Gibbons, on the other hand, exhibit little or no dimorphism and are monogamous. Both sexes compete for mates best able to help in rearing the offspring. Selection for size appears important to females as well as males.

Primatologists note that the dimorphism of chimpanzees, our closest relative, is comparable to our own. While promiscuous breeders, male chimps may compete to share females, but they also cooperate with other members in a troop to share food resources.

Extrapolation to early human societies, however, is a complicated affair. Every primate breeding strategy known is practiced in contemporary human cultures. As anthropologist Helen Fisher notes, "We seem to have borrowed strategies from all the great apes."

Whatever advantage a great size difference like that of gorillas, for example, might have served early hominids, the degree of difference has changed. Anthropologist Nancy M. Tanner of the University of California at Santa Cruz points out that "human selection seems to have been for locomotion, freeing the hands, and for basic communication and cognitive ability." A large size difference wouldn't have been compatible with the cooperative way in which humans coexist.

Might speech, intelligence, and the ability to get along with others in the band have been uppermost in our foremothers' minds when choosing among our potential forefathers? Was she responsible for cutting him down to size?

Such a hypothesis requires broadening to trace the origins of male-female size difference. And as with the other theories, many of the necessary details are open to dispute. In the meantime, men continue to be bigger than women. Whether equal opportunity will ever have the sexes seeing eye-to-eye remains to be seen.

—Bruce Fellman

Why Isn't Life Evenhanded?

Suppose the person you see in the mirror every morning were one day to step into this world. How would that person differ from you? As a glance in the mirror will show, if you are right-handed, your mirror image would be left-handed. Unless you are one of the rare people with transposed organs, your double's heart would be toward the right side of the body.

But there would be more subtle differences as well. When most of the molecular building blocks of life, such as the amino acids that combine to form proteins, are synthesized in the lab, they take forms that are identical in every way but one: They are mirror images of each other. Chemists refer to this duality as handedness. Structurally, almost any amino acid can be either right-handed or left-handed.

But the chemistry of life and the chemistry of the lab are two different things. In life, these small molecules almost always occur in just one of their two mirror images. Though no one knows why, all but one of the amino acids in proteins are left-handed. And so are the main constituents of DNA, the four nucleotides, or bases. It makes sense that they aren't a combination of both. When left-handed amino acids combine, they often form helices—chains that spiral like the threads of a screw. If a right-handed amino acid were among them, it would twist a thread in the opposite direction. But what kept these acids from all being right-handed, as in a looking-glass person?

Some four billion years ago, when the first molecules were forming that would create life, equal numbers of right-handed and left-handed molecules must have been present, because ordinary chemical reactions do not distinguish between the two. In a test tube, about half the molecules that form are right-handed and half are left-handed. Somehow, in nature, the left-handed molecules came out on top.

Perhaps they just got lucky. At some point early on, there may have been a few more left-handed molecules than right-handers. With the odds slightly in their favor, the left-handed molecules might have joined into reproducing groups, creating more left-handers. They simply could have overrun their right-handed competitors.

The other possibility is that some kind of extraordinary chemical reaction gave the left-handed molecules a slight advantage. For instance, polarized light might have decomposed right-handed molecules, leaving an abundance of left-handed molecules to form life. However, polarized light is rare in the natural world. Sunlight reflected off the surface of water becomes weakly polarized, but it is hard to see how this feeble source of light could exert a strong enough effect.

Another idea has to do with where life got its start. Molecules might have come together on the surfaces of clays or crystals such as quartz. Because of their surface geometrics, some of these substances attract only left-handed molecules. But such an advantage could be conferred by other crystals that attract only right-handers. In addition, a crystal like salt attracts both, giving an advantage to neither.

A more exotic suggestion relates the handedness of life to a fundamental handedness of the universe. Until 1957 physicists were sure that the laws of physics were completely indifferent to left and right. Then, to their amazement, they discovered that certain kinds of radioactivity are predominantly left-handed.

Almost immediately the connection was made. What if, in the beginning, this radioactivity gave lefties an advantage, say by breaking up right-handed molecules? Couldn't that account for life's mysterious twist? Despite more than 20 years of trying, no one has conclusively demonstrated such an effect in a laboratory.

In fact, it may take looking far beyond Earth to tell for sure whether left-handed molecules are the rule. If, for example, life that has evolved independently on other planets almost always consists of left-handed building blocks, then they must have an advantage. But even that may not settle the question. We still may not know what that advantage is.

—*Steve Olson*

What Causes
Left-Handedness?

A soldier who holds his shield with his left hand and his weapon with the right, Thomas Carlyle wrote in 1871, betters his chance of living to fight another battle. The combatant who carries his shield with his right hand can't as readily protect the left side of his chest—where his heart lies—from enemy blows. Learned ages ago, this simple rule of survival was passed from one generation of soldiers to the next and, according to the historian, may explain why very few people—about 10 percent of the population—now use the left hand.

Humans are unique in their preference for the right hand. Among other animals studied, individuals favor one foot over another, but in each species, the spilt between right and left is even.

Anthropological studies show that lefties have been woefully outnumbered since the beginnings of recorded history. In artwork dating back 5,000 years, more than 90 percent of the people depicted performing manual tasks use the right hand. A survey of Cro-Magnon hand tracings, which tallied the number of times the left or right hand was outlined, found the same tendency.

A few psychologists still believe, as did Carlyle, that through the ages, cultural pressures—from the rules of war to the dictums of penmanship teachers—have been responsible for hand preference. That doesn't explain, however, why no human society has ever favored the left.

Most psychologists contend that genes account for some of the imbalance, though it isn't clear how hand preference is passed from one generation to the next. Studies of families show that the child of two lefties has about a 50 percent chance of being left-handed. English psychologist Marion Annett at Lanchester Polytechnic suggests left-handedness is the result of two recessive genes that actually code for a lack of handedness. Children receiving two recessive genes would then have an even chance of being right- or left-handed.

It's widely believed that the asymmetrical organization of the brain is tied to handedness. In the mid-1860s, French neurologist Paul Broca found that in right-handers, language is controlled by the left hemisphere of the brain. Broca theorized that in lefties language control would be flipped to the brain's right side. But by the 1970s, neurologists had discovered that the speech center lies in the right hemisphere in just a few of the left-handed—about 15 percent, according to one study. Though scientists still believe there is a connection, they are not sure how language and hand preference are related.

More than 100 years after Broca found a link between brain organization, language, and hand use, Canadian psychologist Paul Bakan at Simon Fraser University proposed that brain damage suffered during fetal development or at birth was responsible for left-handedness. According to his still controversial hypothesis, which is based on work with monkeys, an oxygen deficiency damages left hemisphere cells. To compensate, the developing brain shuffles control of motor skills and other functions to its right side. Bakan further theorizes that though the injury is slight in most left-handers, more extensive injury could account for the disproportionately high numbers of left-handers among people who suffer from dyslexia, stuttering, epilepsy, autism, and other brain-related disorders. However, reorganization of brain functions may turn out to be advantageous for some left-handers: higher than expected numbers of lefties also show up in samples of artists and mathematically gifted children.

Two psychologists, Michael Corballis at the University of Auckland in New Zealand and Michael Morgan at the London School of Economics in England, suggest that at its most basic level, nature itself prefers asymmetry. This universal imbalance could account for biological directional preferences that range from the rotational movement of single-celled organisms—all the members of a species tend to swim in the same direction—to the spiraling of a snail shell to the organization of the human brain.

It seems, say the psychologists, that genes and other factors play secondary roles reinforcing or disrupting an underlying influence that, in the case of language and hand skills, steers the brain development of a fetus toward left hemisphere specialization. Corballis and Morgan suggest that this unidentified steering influence, or "left-shift factor," is found in the fluid of the unfertilized human egg. But occasionally the left-shift factor, whose presence in the fluid is probably genetically influenced, is absent or overpowered by, say, other genes or an imbalance of hormones. Then brain organization in the developing fetus may deviate from the standard pattern of asymmetry—left-handedness is one possible outcome.

Why would nature impose a left shift on brain development? An even more fundamental handedness of the universe may be responsible. According to physics experiments performed in 1957, some subatomic processes, such as certain forms of radioactivity, favor the left. A link between nature's fondness for the left and hand preference remains to be proven. But if such a link exists, then lefties, bucking a near-universal trend, may be the ultimate nonconformists.

—*Mark Bello*

Why Do We Cry Tears?

Your dog dies. You win the beauty pageant. You break up with someone you love. Your daughter gets married. You lose your job. Your best friend has a serious accident.

How do you handle such stressful episodes? Chances are you cry. Shedding tears seems to bring about a terrific emotional release. Why? No one can say for sure.

Humans apparently shed a variety of tears. There are the tears we secrete all the time, those to help keep our eyes properly moistened. Then there are irritant tears, the kind we spill when peeling an onion or coping with smog. Finally, there are emotional tears, the stuff we trickle in response to grief, joy, frustration, or other stresses.

Curiously, of all the earthly creatures, only humans seem to shed emotional tears. That makes such tearing a late evolutionary development. Tears of emotional stress also appear relatively late in infant development. Unhappy newborns often cry tearlessly until they are several days old or even until weeks after they are born. But challenge them with an eye irritant, and they can spill tears at birth.

Charles Darwin proposed what appears to be the first scientific theory—or, rather, nontheory—to explain emotional tearing. According to Darwin, as expressed in his book *The Expression of the Emotions in Man and Animals,* published in 1873, it was the total act of crying that relieved suffering and made people feel better—not the secretion of tears, which, he contended, was an incidental and in itself purposeless accompaniment to the catharsis.

About three decades ago another theory surfaced, this one promulgated by Ashley Montagu, the noted anthropologist. According to Montagu, the tears that went with sobbing did indeed have survival value in that they helped to protect us against disease. He argued that sobbing—gasping and convulsive catching of breath—dried out nose and throat membranes, thereby increasing vulnerability to bacterial invasion. Tears, which also drain into the nasal passages, served to offset this tendency towards dryness.

Both these theories leave plenty of room for counterargument. The big hitch in Darwin's theory is this: It's hard to imagine evolution promoting the development of a purposeless function. Coming from Darwin, the very person who saw purposeful natural selection as the fuel of evolution, such a proposition smacks of out-and-out heresy. Montagu's theory, meanwhile, doesn't seem to take into account the fact that many people cry silently. Why is it that they shed tears even when there's no perceptible change in breathing—no sobbing to dry out the nasal passages? And why doesn't the heavy breathing that accompanies strenuous exercise such as running or swimming induce the athlete's tears?

Dissatisfied with both theories, William H. Frey II, a biochemist and director of the psychiatry research laboratories at the St. Paul-Ramsey Medical Center in Minnesota, has recently suggested another hypothesis. He proposes that tears may help to rid the body of chemicals produced by emotional stress. According to Frey, when we need relief, we may literally "cry it out." He argues that all other excretory functions—urinating, sweating, exhaling, defecating—are involved in removing excess or toxic products from the body. Why shouldn't the same hold true for emotional tears?

On the face of it, Frey's theory seems most plausible. But for now it remains purely speculative, as do the postulates of Darwin and Montagu. None has either been verified or refuted in the lab. Frey's hypothesis, however, seems most amenable to experiment. As such, the Minnesota investigator recently began a series of trials aimed at testing its validity. One thing he's doing is having volunteers watch tear-jerkers; his favorite is *The Champ,* a movie about a down-and-out boxer and a little boy. He compares these emotion-provoked tears with irritant tears collected from the same subjects while they peel onions. If Frey's theory has merit, then there should be a significant difference in the chemistry of these two varieties of tears. Results from a group of over 80 subjects suggest there are. Emotional tears contain a greater concentration of protein than do irritant tears.

But there's no telling what, if anything, this protein difference means. Are there differences in protein kind as well as quantity? Are there specific proteins associated with emotion? If so, how do they relate to the hormones or other agents that mediate our emotions? Is there a specific substance associated with each emotion? Is there one agent, for instance, that makes us feel anger, another, elation, and yet a third, grief? And are these differences *all* reflected in tears?

—*Ben Patrusky*

The Nose Knows, But Science Doesn't

Noses have it pretty hard. Boxers flatten them. Doctors rearrange them. People make jokes about their unflattering characteristics. Worst of all, when it comes to smell, no one really understands them.

Despite the nose's conspicuous presence, its workings are subtle. Smell, or olfaction, is a chemosense, relying on specialized interactions between chemicals and nerve endings. When a rose, for example, is sniffed, odor molecules are carried by the rising airstream to the top of the nasal cavity, just behind the bridge of the nose, where the tips of tens of millions of olfactory nerve cells are clustered in the mucous lining. The molecules somehow trigger the nerve endings, which carry the message to the olfactory lobes of the brain. Because smell information then travels to other regions of the brain, the scent of a rose can elicit not only a pleasurable sensation but emotions and memories as well.

Though just how odors stimulate the nerves is unknown, scientists do know that our sense of smell is surprisingly keen, capable of distinguishing up to tens of thousands of chemical odors. The laboratory task of isolating the components of an odor is far from simple. Tobacco smoke, for example, is made up of several thousand different chemicals. Moreover, smell researchers must grapple with the problem of what to call the different odors that the nose detects. People generally refer to smells by their sources or associations. Descriptions such as "like a wet dog" or "like my elementary school" may convey perceptions but are vastly inadequate for labeling the chemistry involved.

To further complicate research, olfaction is connected to other sensations. Besides olfactory nerves, the nasal cavity contains pain-sensitive nerves that perceive sensations such as the kick in ammonia or the burning in chili peppers. Smell also intertwines with taste to create flavor. A coffee drinker holding his nose while sipping would taste only the bitter in his brew, for taste receptors generally appear limited to bitter, salty, sour, and sweet. The sense of smell is ten thousand times more sensitive than taste and makes subtle distinctions between lemon, chocolate, and many more flavors.

So how does the nose manage this sophisticated discrimination? Lack of evidence hasn't kept scientists from speculating. One idea is that every odor molecule vibrates at its own frequency, creating patterns of disturbance in the air similar to the wave patterns produced by sound. According to this theory, the nerves act as receivers for the unique vibrations of every odor molecule. The scheme requires no direct contact between the molecule and the nerve cell.

Another suggestion is that primary odors, equivalent to the primary colors of vision, underlie all smells and are detected by receptor sites on the olfactory nerves. Different combinations of about 30 basic smells, with labels such as malty, minty, and musky, could form an infinite number of odors.

Other scientists think that each smell is its own primary

smell. They believe the olfactory nerve endings have specific receptor proteins that bind to each of the chemicals people can sense. This theory, however, calls for thousands of different proteins—none of which has been found.

"The science of smell is so empirical," says Robert Gesteland, a neurobiologist at Northwestern University, "there's no predictive base for experiments." Unlike the senses of sight, touch, and hearing, olfaction studies have attracted only a small share of scientific interest. That may change. Researchers hope that unraveling the mystery of smell will advance our understanding of the brain. Smell research promises to help physicians better diagnose the smell and taste disorders that affect two million Americans. And in the future, with enough known about smell, it might be possible to endow strange, unappealing but nutritious foods with more familiar odors, perhaps expanding the world's food supply. For the moment, however, what the nose knows it isn't revealing.

—*Adriana Reyneri*

What Makes Teeth Sprout?

As infants, we suffered the pangs of teething, and as children, visions of the tooth fairy gave way to the realities of the dentist and the orthodontist. With the crowning of our wisdom teeth, 20 years or so have passed, and the cycle has started again in another generation.

Reasons why teeth come in and fall out are slowly emerging. But no one in dental science can certify what forces push teeth through the gums or identify the timing mechanism that schedules their eruptions.

Actually, our teeth undergo constant change throughout our lives, from the first trimester of prenatal life until our last bite. Six weeks after conception, tooth germ cells are developing from the primitive tissues of the mouth. At birth, deciduous or baby teeth are obvious in X rays, and some of their permanent successors are steadily growing. By age three, the roots of the 20 deciduous teeth are fully formed. At around six years, the first of the 28 permanent teeth, excluding wisdom teeth, begin to emerge.

The growing caps of permanent teeth, which put pressure on the roots of deciduous teeth, have moved toward the surface of the gums with help from a basic principle of bone biology: where bone is under pressure, it disintegrates and is taken up, or resorbed, in a process called osteoclastic action. The roots of the shedding tooth disappear under pressure, while the cap of the emerging tooth is protected by calcium and other minerals that are hardening its crown. (Osteoclastic action, in fact, is what orthodontists rely on when they reposition teeth under pressure of wire and rubber bands.)

Just as there is an abundance of X-rays and anatomical specimens showing teeth in every stage of development, there are plenty of theories to explain the forces of eruption. Off the cuff, root growth seems a likely theory. But roots must push against something, and the principle of osteoclastic action does not permit bone to act as a fixed base. In experiments with rodents, when the erupting tooth is prevented from moving, the roots keep growing and the bony sockets in which teeth are positioned are resorbed where the roots push. Moreover, rodent experiments show that if the tooth is severed in half while erupting from the gums, the top half will emerge and shed anyway.

Others have postulated the existence of another ligament, in addition to the periodontal ligament that surrounds teeth in their sockets. A "cushioned hammock ligament" may straddle the bottom of the bony socket like a sling and act as a fixed base for the roots. But dental anatomists argue that no such function is apparent.

Perhaps vascular pressure in the surrounding tissue is responsible for tooth eruption. Experimental measurement of pressures in unerupted teeth of dogs indicates enough pressure to cause movement. But this theory suffers from the fact that the tops of teeth emerge in spite of separation from their roots and the surrounding tissue.

Perhaps the most intriguing explanation for both timing and emergence is one that is linked to the production of collagen—a protein and chief constituent of connective tissue. Collagen is produced in the cells of the periodontal ligament. The theory is that when ligament cells called myofibroblasts shrink, the collagen fibers connecting them are pulled tight, which creates contractile forces that squeeze teeth along in their paths. As yet, no one has demonstrated the presence of myofibroblasts in the periodontal ligament. But when scientists interfere with collagen production in the ligament, tooth eruption is retarded, roots buckle, and the bony socket is resorbed.

Teeth, like bone, are made from tissues and may be responding to hormonal signals, says Michael Roberts, a dentist at the National Institute for Dental Research. "But nobody has been able to demonstrate a hormone or a receptor site," he says. Hormonal stimulation could explain timing and link our genes to tooth development.

The genetic research that has been done on the emergence of teeth leads scientists to believe many genes might be involved. Tooth deformities, timing errors, and eruption failures accompany quite a few genetic diseases. The sheer number of genetic targets for study is immense.

Even without hard answers to the riddles of emergence and timing, the tooth fairy will be as busy as ever. Perhaps the biggest question is whether she can keep up with the pressures of inflation.

—Allen J. Seeber

What Causes Aging?

Throughout its history, humankind has never quit dreaming the grandest of dreams: finding a way to slow or halt the process of aging. The latest seekers of the legendary Fountain of Youth are, logically, biologists; for the still undisclosed place, if there be one, will likely be found within the body. Only by locating and deciphering the machinery of senescence, say these latter-day Ponce de Leons, is there hope of discovering if we have any chance of prolonging life.

Perhaps we don't. Every species seems to have a specific life-span or maximum age. Contrary to what many people believe, progress in medicine has not increased this limit. It remains what it has always been: the biblical four score and ten. What medicine has done is enable more people to live out a fuller measure of their allotted years.

But it has done nothing to stave off the manifestations of aging: graying hair, loosening teeth, weakening bones and muscles, increasing susceptibility to disease, wrinkling of skin, menopause. Bernard Strehler of the University of Southern California estimates that after the age of 25 or 30, the body relentlessly loses functional capacity at the rate of about one percent annually. What causes this inexorable decline? And is it unalterable?

There is no shortage of aging theories. And no single hypothesis attempts to explain all of the characteristics of senescence. There are two broad areas of consideration: Some scientists concentrate on the aging processes at the cellular level, and others study senescence in terms of biological systems.

Favored among the cellular-aging theories are those centering on the genetic machinery—the DNA and its associated protein-making apparatus. One idea is that the machinery within each cell wears out, gradually losing its capacity for self-repair and resulting in deterioration. Another proposal has to do with redundant genes. This "iterative" gene theory suggests that a mechanism exists for casting off a faulty component and moving in a replacement gene. Longer-lived species presumably have more reserve DNA than short-lived species. Ultimately, all creatures exhaust these reserves. A third theory argues for the existence of an "aging" gene, maintaining that senescence is written into the genetic script at birth as part of the normal developmental program. The aging gene, or genes, "switch on" at a preprogrammed time. Once activated, the genes act to slow down or halt manufacturing processes crucial to the life of the cell.

Meanwhile, other cell biologists use the "free radical" theory to explain aging as a product of deterioration in the cell's energy-processing center rather than the genetic machinery. Free radicals are unstable atoms, transient by-products of the bucket-brigade process by which the cell converts food-stuffs to fuel. Gerontologist Alex Comfort describes the free radical as "a highly reactive chemical that will combine with anything around." Unless quickly mopped up, free radicals can do a heap of damage to cell structures. There are clean-up enzymes to do just that, but with the passage of time, those enzyme levels diminish.

Not all aging specialists look at senescence exclusively as a result of the aging of individual cells. They describe aging as a consequence of disruptions in various regulatory mechanisms operating within the whole organism. One such theory, for instance, implicates the immune system, which manufactures antibodies against foreign invaders. In time, according to the autoimmune hypothesis, the body grows allergic to itself. That could happen because cells undergo surface changes or because "sloppiness" in the manufacture of antibodies blurs their ability to distinguish self from nonself. In either case, the antibodies attack the body's own cells, producing the ravages of age.

Another organismic theory suggests aging may be programmed in the brain. This "hormonal clock" hypothesis proposes that at a set time the pituitary gland releases a "death" hormone, which triggers a host of age-associated disruptions, including the creation of excess free radicals and the accumulation of errors in the genetic machinery.

Probably several of these theories will turn out to be partially true, with some mechanisms acting jointly to produce senescence. If so, the age-old dream of extending life-span will probably remain an impossible dream. But all is not lost. Solving the mystery of aging could help preserve the health associated with youth to the preprogrammed maximum age, thus making it possible to grow old with more than grace.

—*Ben Patrusky*

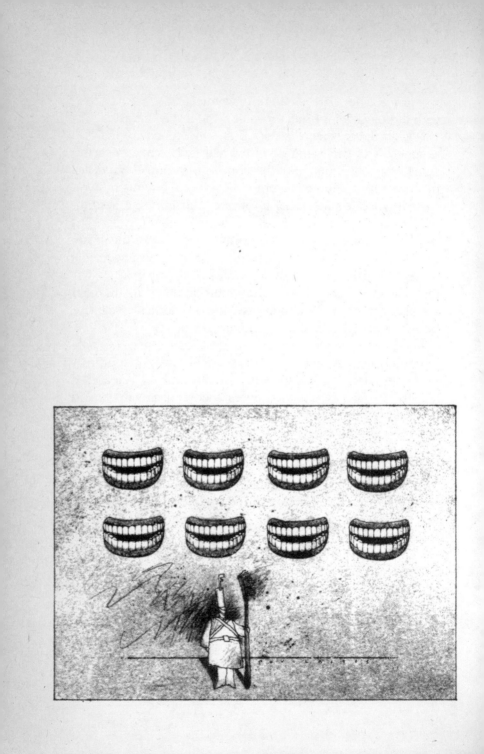

Part III
HEALTH

Why Do Women Live Longer Than Men?

Life expectancy is at an all-time high, trumpeted a brochure released not long ago by the Metropolitan Life Insurance Company. But women have a clear statistical edge. According to the latest data collected by the National Center for Health Statistics, a baby boy born today can attain the ripe old age of 71.2, while a baby girl can reach 78.2. This seven-year disparity has yet to be explained by the medical community.

A female's ability to outlive a male was noted as early as the 17th century, but it wasn't until this century that the difference in life-span became a gaping chasm. In 1980, there were 131 women for every 100 men between the ages of 65 and 74, and by the year 2000, predictions run to 150 women for every 100 men in that age group. "The gap is widening all the time," notes Edward Schneider, associate director of the National Institute on Aging. "And we really don't know why."

While both men and women are living longer, women seem to have benefited most from the elimination of infectious diseases as the top killers. Because the gap in life expectancy is largest in industrialized societies, women may have some kind of natural resistance to the so-called diseases of civilization, like alcoholism and cardiovascular disorders.

But just how genetic and behavioral differences might affect life-spans has been the subject of few scientific inquiries. Erdman Palmore of Duke University estimates that "about half the greater longevity of women is due to genetic differences and about half is due to differences in life-style, such as less hazardous occupations and more careful driving, which produce lower accident rates, and less cigarette smoking, which produces less lung cancer and cardiovascular disease."

While most medical researchers agree that the phenomenon is the result of a combination of factors, not enough is known about them to explain the gap. An oft repeated refrain is that as more women enter the work force and take up smoking, the gap will close. Yet ever since World War II women have been working and smoking in greater numbers and are consistently increasing their edge. As the King of Siam would say, "It's a puzzlement."

Some think that women have reaped the benefits of the Industrial Revolution in the form of greatly improved medical techniques for childbearing. At the turn of the century, less than five percent of American babies were born in hospitals. By 1979, just about all U.S. births occurred in hospitals with a decline in maternal deaths. Men, on the other hand, continue to suffer the negative effects of financial responsibility and stress, which include a higher incidence of alcoholism and liver disease, of suicide and today's Public Enemy Number One: heart disease.

Another prevalent theory is that women have more contact with the medical care system. "Perhaps women are taught to take better care of themselves and to deal with their health problems when they occur," suggests medical sociologist Marcia Ory. "However, national statistics indicate that while women report more illnesses than men, they have a lower incidence of life-threatening diseases."

In addition to taking advantage of expanded health services, it could be that a woman's life-style is healthier than a man's. On average, a woman doesn't drink or smoke as much as a man, she sleeps more and has a wider social network to relieve stress and combat loneliness. Others theorize that women have less exposure to physical and social stresses in the work environment, or different and healthier ways of coping with stress.

M. William Voss, director of the Department of Geriatrics and Gerontology at the University of Maryland, believes there's a strong psychological component to the higher mortality of males. "Many men," he notes, "seem to lose their reason for living when they retire. I've observed healthy farmers who've worked hard all their lives retire and move to town. They sit rocking on their front porches and a few years later have a heart attack and die."

A National Institute on Aging study indicates that widowed men fare worse after the death of a spouse than women do, suggesting that men's health may also be more severely affected by emotional blows. Recognition of these factors may help men better their life expectancies, but there's a growing body of evidence that suggests that women would still have an advantage in the battle for survival.

"Throughout the mammalian kingdom, there's no contest," asserts Estelle Ramey, professor of physiology and biophysics at Georgetown University School of Medicine. "Females appear to have the natural advantage from the moment of conception."

Ramey attributes this advantage to an inherent sex-linked resistance to life-threatening disease. Her research with Peter Ramwell indicates that a female's hormones work to give her a more efficient immune system. The different effects of the female hormone estrogen and the male testosterone point strongly to estrogen as a protective factor against heart disease. During their childbearing years, when estrogen levels are high, women have a significantly lower incidence of cardiovascular disease than men of the same age. And studies on the use of postmenopausal estrogen indicate that estrogen can aid in the prevention of heart disease.

But just how estrogen beefs up the immune system isn't fully understood. "We need much more research," says Ramey. "People are not notable for acting rationally about sex differences, and the denial of women's longitudinal advantages may have inhibited research in the past."

"There are many promising signs that more research is taking sex differences into account," says Ory. "We're beginning to understand that each individual's body chemistry works differently and that biology interacts with social factors."

But the total picture remains elusive. And the gap between life expectancies continues to widen.

—Sue Hoover Epstein

Decade Of The Killer Brain Infection

A group of neurologists once speculated that the rise of the Nazi party was partly due to a brain infection known to cause extreme psychiatric disturbances, including pathological violence. The affliction proposed was encephalitis lethargica, and quite apart from any role in Hitler's scheme, it remains one of the most bizarre diseases ever known.

Although encephalitis lethargica probably also broke out in China, it was first described in Vienna during the winter of 1916-17, marking a worldwide epidemic. Spreading through Europe and America, the disease raged for 10 years, striking five million people and killing half a million before disappearing in 1927, as suddenly and mysteriously as it had arrived.

Even the early symptoms of the disease were much more varied than the somnolence and stupor that led to its better-known name, sleeping sickness. (The African sleeping sickness, trypanosomiasis, is unrelated to the encephalitic disease.) In addition to headache and fever, victims of encephalitis lethargica suffered muscular spasms, seizures, difficulty in swallowing, and uncontrollable eye movements. Young adults and children were most often hit: some became psychotic, acting with a manic destructiveness that in a few instances led to murder. Even after apparently recovering, many were plagued by involuntary movements and compulsive behaviors—traits of Parkinson's syndrome, only far more extreme than those typically seen.

A third of the sufferers died, either in a deep coma or a state of tormented sleeplessness that was beyond sedation. It seemed, in fact, as though several different diseases had broken out simultaneously. "Epidemic schizophrenia" and "atypical poliomyelitis" were two of many names given the condition. The person who discerned its true nature was Austrian neurologist and pioneer aviator Baron Constantin von Economo. Studying the brains of deceased victims, he noticed a common pattern of damage. By injecting contaminated brain tissue into monkeys, he found that the disease was contagious and concluded that it was caused by an invisible microbe—probably a virus.

Since then, several other types of encephalitis have been conclusively traced to viruses, some of them spread by mosquitoes in the summer. But the precise identity of the agent that struck terror into the hearts of Europeans 70 years ago remains as elusive as ever.

A few victims of the pandemic are alive today, and their plight illustrates another tragic feature of encephalitis lethargica. These people never again experienced a normal life. They survived their disease but in a zombielike state. Neurologist Oliver Sacks first examined these survivors at a suburban hospital in New York, where they sat "motionless and speechless all day in their chairs, totally lacking energy, impetus, initiative, motive, appetite, affect or desire; they registered what went on about them without active attention, and with profound indifference." Lack of will, writes Sacks, "forms the empty heart of such states."

In 1969, Sacks began testing over 200 of these patients with L-dopa, the precursor of the natural brain chemical dopamine. The treatments followed the discovery that L-dopa could be used successfully on patients with ordinary Parkinson's syndrome, which is linked to dopamine deficiency. Sacks's early results were promising. One patient began to speak, write, and walk—as well as recover her charm, intelligence, and humor—for the first time in over two decades. But as with Parkinsonism, L-dopa proved less miraculous over time. Another patient's improvement was extremely brief. After an initial elation, she seemed to find intolerable the discovery that half a century of her life had been lost. Her response to L-dopa rapidly diminished.

Textbooks tend to compare the encephalitis lethargica episode with infections like scarlet fever, which disappeared partly because the disease bacterium seemed to decline in virulence. Perhaps the sleeping sickness virus did likewise, only more dramatically. Then again, there are occasional reports of infections similar to encephalitis lethargica. In 1983 doctors at Addenbrooke's Hospital in Cambridge, England, reported three suspicious cases in children. Some experts think that the disease has always been around, that it is endemic throughout the world. If so, the mysterious malady prompts two further questions. What, in the winter of 1916-17, turned a sporadic infection into a harrowing, epidemic killer? And might it not happen again?

—*Bernard Dixon*

Why Do People Stutter?

It's been more than two millennia since Aristotle proposed that stuttering resulted from a bad mix of the body's four humors and over a century since a Prussian surgeon vainly tried to cure stutterers by snipping off portions of their tongues. Fifty years ago at the University of Iowa, Wendell Johnson, a stutterer turned specialist in his own handicap, began the first systematic study of the affliction that affects one or two people out of every hundred. But science has yet to determine a cause for stuttering.

For some, stuttering is transient—starting between the ages of two and seven and disappearing by adulthood. For one stutterer in five, however, the impediment continues. Therapy helps increase fluency but usually at the expense of slowing speech.

Characteristically, a stutterer's speech is studded with rapid-fire repetitions and sudden pauses. Because voice production requires coordination of the brain, breathing apparatus, vocal cords, tongue, jaw, and lips, physiologists have tried to trace the problem to malformed parts.

Some investigators thought stutterers merely lacked enough wind to push their words out. Others viewed stutters as small seizures, labeling stutterers as epileptics. Still others conjectured the disorder resulted from a fight between the brain's two hemispheres for the control of speech. Despite such theories, the brains and lungs of the majority of stutterers appear no different from those of nonstutterers. In fact, the occasional slips of normally fluent people usually occur at the beginning of sentences and phrases or after a breath—the same roadblocks that detain stutterers.

Stutterers sometimes lose their affliction—when they sing, whisper, speak in unison, form their words while inhaling, or talk without hearing themselves. Under stress, however, the problem worsens. Many a stutterer becomes verbally impotent when speaking in public, talking on the telephone, or making introductions. The schoolchild who can recite the pledge of allegiance perfectly with his classmates may find himself tongue-tied when called on by a teacher.

Psychologists by the dozens have traced the disorder to unresolved conflicts caused by demanding teachers or perfectionist parents. But an equal number cited stutterers who came from nurturing environments. Freudian thinkers believed hostility and anxiety were expressed through disturbed speech. Yet psychotherapy fails to cure most cases of stuttering, and countless psychological profiles reveal that stutterers are no more neurotic than the rest of the population.

Believing himself neither mentally nor physically deficient, Johnson proposed that stuttering begins in the ear of the listener: If called a stutterer, a child will learn to stutter, reasoned Johnson. But he never proved it.

"Stutterers have been poked and probed and tested in almost every conceivable way," says Martin Adams, a speech-language pathologist from the University of Houston. "And it has become increasingly obvious that stuttering is not due to a single cause."

Studies show that stuttering runs in families, affecting four times as many males as females. What's more, the evidence suggests it is not the stuttering that is inherited but the predisposition to become a stutterer.

"The existence of the predisposition does not sentence the child to becoming a stutterer," says Frances Freeman, a speech physiologist from the University of Texas at Dallas. Some form of real or perceived stress may be needed to trigger the disorder. Some stutterers may inherit slow word retrieval functions or poor basic speech coordination. In these cases, something as simple as parents who talk rapidly may trigger stuttering in the child.

Thus, the appearance of a stutter seems weighted by two complex factors—genes and environment. But the most likely combinations of inherited and environmental influences that give rise to the disorder elude scientists. "We still don't know what causes stuttering, and anybody who professes to know is barking at the moon," says Adams.

—Lynn J. Cave

How Does Fluoride Fight Decay?

Every time you pop a peppermint into your mouth, millions of bacteria rally their forces to begin a process of rot and decay. The microbes use sugar to produce acids that dissolve calcium and phosphate in tooth enamel, weakening its structure. Eventually bacteria break through the surface and a cavity begins.

Unless, of course, the process is interrupted. By the 1930s, researchers had noted that children were less likely to develop cavities if they grew up in areas with naturally fluoridated water. Toward the end of World War II, Grand Rapids, Michigan, became a scientific guinea pig as the first community to artificially fluoridate its drinking water. Now almost 50 years later, half the people in America drink fluoridated water and nearly all the toothpaste they use contains the well-advertised decay fighter. Studies show that fluoridated water reduces the development of cavities by an average of 60 percent and dentists routinely recommend brushing with fluoride toothpastes. So although the John Birch Society still claims it's all a communist plot, most scientists are pro-fluoride. But they can't prove how it fights the damage done by bacteria.

Early research on fluoride focused on the developing teeth of young children, and it pretty well established that fluoride strengthens the enamel's crystal structure. When fluoride from drinking water passes through the bloodstream to children's unerupted teeth, charged particles, or ions, of fluoride replace other, larger ions in the developing enamel. The fluoride ions pull the molecules attached to them into a tighter, less soluble crystal lattice that may make teeth more resistant to bacterial acids.

But fluoride has more than one way of fighting decay, in adults as well as children. A cavity, at any age, takes up to eight years to develop. Initially, the calcium and phosphate that are dissolved by bacterial acids get absorbed by plaque, the sticky film of food particles and bacteria that coats teeth. However, each time bacteria run out of sugar, the saliva's acidity drops, and in a natural healing process called remineralization, the damaged enamel begins to reabsorb the lost minerals. In what may be its most important role, fluoride seems to accelerate the uptake of the minerals by scarred enamel, though no one is sure how. Proteins produced by the salivary glands inhibit remineralization, and it may be that fluoride squelches the action of these proteins. Or fluoride may aid remineralization by converting more calcium and phosphate into a form that the enamel can absorb. Fluoride taken up also strengthens enamel crystal.

Fluoride also may attack the enemy directly, by upsetting the metabolism of bacteria and slowing acid production. But scientists don't know at what stages in the metabolic process the fluoride goes to work. It may inhibit the synthesis of necessary enzymes and their movement in bacteria. This biochemical interference, however, only happens when fluoride concentrations are as high as those found in toothpastes and rinses.

Fluoridated water, on the other hand, might protect teeth in yet another way. Some researchers think that children exposed to fluoride from birth may develop molars with shallower pits and fissures, leaving less room on the smooth surfaces for food to build up and for bacteria to grow. Animals consuming fluoride tend to produce smaller, smoother teeth, which scientists attribute to thinner layers of enamel and the underlying structure, dentin. No one knows, however, why the layers are thinner, whether fluoride slows the rate at which calcium and phosphate minerals accumulate, or if it somehow upsets the metabolism of the enamel-producing cells.

While fluoridated water is the source of a dramatic, overall reduction in cavities, it does not affect all age and demographic groups to the same degree. Knowing how fluoride works could help target treatments more effectively. And even children from unfluoridated areas, who soak their teeth in highly concentrated fluoride at the dentist's every six months, develop 20 to 40 percent fewer cavities. According to James Bawden at the University of North Carolina, those rates are too good to make sense. The fluoride in the saliva is gone with the next glass of unfluoridated water. And what fluoride is absorbed by the enamel lasts a few weeks at best. We know the treatment works. But, as with much of fluoride research, says Bawden, "there's a lot of evidence that needs to be nailed down before we explain how."

—Diana Morgan

How Does Smoke Kill?

The fire in room 404 started the way many do, with a cigarette igniting an upholstered chair. The occupants of the room escaped unharmed, but 12 other guests at Houston's Westchase Hilton Hotel died from the fire that night, even though flames never spread beyond room 404. Like almost 6,000 Americans who die each year in fires, they were victims of smoke, not flames.

In fact, 80 percent of fire deaths are the result of smoke inhalation, but researchers have a lot of questions about what it is in smoke that kills. "We don't know to this day and we never will, exactly, what killed those people in Houston," says Gordon Vickery, former chief of the U.S. Fire Administration and president of the Foundation for Fire Safety, headquartered in Arlington, Virginia.

The puzzle is that carbon monoxide, the supposed villain, turnes out not to be. At least not alone. Carbon monoxide is an odorless, colorless gas given off by nearly everything that burns; when inhaled, it binds with red blood cells, preventing them from carrying vital oxygen to the body's cells. In the Houston fire only two of the 12 victims had anything close to lethal doses of carbon monoxide in their blood. Something else contributed to their deaths.

One suspect is hydrogen cyanide, more familiar as the killer in gas chambers. Hydrogen cyanide is produced when nitrogen-containing materials such as wool—or modern building materials such as polyurethane and nylon—burn. An investigation of the Houston fire by the Foundation for Fire Safety linked cyanide in the victims' bloodstreams to nylon carpets and blankets and polyurethane carpet padding and chair cushions. Although cyanide levels were above normal in all the Houston victims, they reached clearly lethal levels in only two, both small children.

Another suspect—really a suspected accomplice—is hydrochloric acid, emitted by burning or smoldering vinyl. When inhaled, this very strong acid corrodes lung and throat tissue. It may be the combined effects of these two gases, and possibly other, less well identified substances, that actually cause smoke-inhalation deaths.

Researchers are frustrated by the lack of hard data on the relative roles of the various toxicants in fire deaths and injuries. For one thing. "We haven't been doing a very good autopsy analysis of fire victims," says Merritt Birky, director of research for the Foundation for Fire Safety. Medical examiners routinely check for carbon monoxide in fire victims but rarely go further. The Houston investigation, which made cyanide measurements, was an exception. "Coroners frequently do not have the resources to do these analyses," Birky says, for they are much more difficult and costly. Except for a handful of projects scattered around the country, there is no routine system for investigating fire scenes for toxic gases or toxic-producing materials.

Just as serious, according to Barbara Chernov Levin, head of fire toxicology research at the National Bureau of Standards, is that there is no system for following up fire victims to check for delayed effects of toxic gases. "It's entirely possible that it takes a few days for pulmonary damage to show itself." Little is known about how serious or wide-spread these aftereffects are.

Levin points out, too, that evaluating what evidence we do have is difficult because of some fundamental unknowns. It's still not clear what the lethal level is, she says, for cyanide or even carbon monoxide, especially in combination with other gases. Individual variations can confuse the issue. Such factors as heart disease or the use of alcohol or medications may make some people more susceptible than others to toxic gases. And even "the panic itself may cause metabolic changes in people trapped in a fire that affect vulnerability."

Researchers hope to answer these questions so that steps can be taken to save some of the 6,000 lives lost each year to smoke. But until then, ignorance will be reflected in inaction: "In America today," Gordon Vickery says, "with hundreds of fire codes and standards, not one includes a rating for toxicity of building materials."

—*Stephen Budiansky*

Why Do Warts Disappear?

Oh, lordy, warts are ugly. They're like tactile swamp gas. They're like that big day when the junior miss contestant from Jefferson County wakes with a zit on her chin. They're like the end of life as we know it.

And is it any comfort knowing that *Verruca vulgaris,* these common warts that infest the body, are caused by viruses? Warts don't come from playing with frogs, from cow udders, egg whites, or certain anti-social behavior. Instead, at least 15 papilloma viruses have replaced dozens of folkloric legends. The viruses are responsible for many kinds of warts, including plantar, butcher's, venereal, and the common warts that appear most often on the hands and knees of children. Scientists know a lot about warts, but not why they disappear.

All warts will disappear, eventually, assuming one's immune system functions properly, that the body's resistance is normal. It may take weeks or even years from the time the wart first appears, but it will leave without a trace if left alone.

Common warts get started when the microscopic virus infects the epidermis, causing abnormal cell proliferation.

The wart grows downward until it hits the dermis. On the skin's surface, most warts spread from about one-quarter to one-third of an inch across with a rough but flat surface. Remarkably, when the wart leaves, the skin simply closes back over, like a fissure after an earthquake. There is no scab, no scar; the flesh is smooth again.

"What actually makes the wart disappear is a question," says Doug Lowy of the National Institutes of Health. "There is a suggestion that immunology plays a role." The body could rid itself of the wart if it became immune to the virus, perhaps by generating antibodies. In laboratory studies, a wart from a subject's foot was ground up and then injected as an antigen to produce an antibody against his other warts. But because wart viruses differ, the antigen was successful only in removing some kinds of warts.

One obstacle to laboratory research on warts is a lack of subjects for the experiments. The human wart viruses won't infect the common laboratory rat and even most humans are immune by age 20. Ideally, scientists would like to find an all-purpose antigen effective against a range of viruses.

Another theory as to why some warts disappear is that they are shed along with skin cells. The epidermis, or upper layer of the skin, is constantly shedding. For a wart to stay put, it must reinfect the new skin cells.

Without knowing why warts disappear, the success of current treatments depends upon somehow destroying the virus for each individual wart. Unfortunately, the viruses can spread from person to person and from one part of the body to another. The old home remedy of picking at a wart may produce more. Even the standard medical remedies such as burning out the wart with acids, freezing the tissue with liquid nitrogen or solid carbon dioxide, or drying out the wart with electrodesiccation can't guarantee the elimination of viruses. If any viruses are missed, they will wait until conditions are right for growing.

Curiously, less direct methods of attack have been successful. "There have been controlled trials with hypnotism in which the warts have gone away," Lowy reports. "Many people use autosuggestion and have their warts disappear. And there are lots of instances of somebody being told that his warts will go away next month, and the warts go away next month." While science can show that imagination doesn't cause warts, imagination may play a role in their cure.

Then there are a host of other remedies that, if successful, defy explanation. For years people have sworn by dandelion juice, cornbread crumbs, tree saps, castor oil, and onion juice as wart ointments. Some say the only way to get rid of warts is to rub them with dried peas, corn kernels, white beans, or new pennies. If that doesn't work, there are treatments like rubbing snails on warts, killing moles over warts, washing warts in pig or frog blood, tying slugs to them, or even flinging a dead black cat into a cemetery at midnight. "Why some treatments work." Lowy says, "is really not clear."

My mother always said that the only way to get rid of warts is to rub your finger on them when church bells are ringing, saying aloud. "The church bells are ringing for my warts, the church bells are ringing for my warts." My mother doesn't have warts. She has beauty marks.

Of course, this is the best way to get rid of warts. Call them something else.

—*Carollyn James*

What Causes Crib Death?

When Marie Valdeś-Dapena went to work in the Philadelphia Medical Examiner's office in 1957, she was shocked to learn that many seemingly healthy babies were dying, suddenly, without cause. The victims left behind frustrated physicians, stunned, grieving parents, and a confused trail of guilt. "Only the coroners knew the magnitude of the problem," says Valdeś-Dapena, now of the department of pathology of the University of Miami School of Medicine, "and even they were saying there was no explanation."

Sudden Infant Death Syndrome, as the problem later would be called, is still unexplained. It kills 7,000 U.S. children each year and is the largest cause of death in infants between the ages of one month and one year. Though no one knows exactly what goes wrong, researchers see the problem, in its broadest sense, as respiratory: For some reason, these babies stop breathing, probably during sleep, and don't begin again.

Most experts believe that an abnormal respiratory control mechanism is responsible for a sequence of events leading to death. According to Alfred Steinschneider, president of the American SIDS Institute in Atlanta, Georgia, the abnormality may be apparent in the first few days of life or even prenatally. Steinschneider reported in 1972 that some infants who later became SIDS victims were subject to abnormally long periods of arrested breathing during sleep. In some adults this same problem had been linked to "underventilation"—an inadequate exchange of carbon dioxide and oxygen in the lungs that diminishes the body's oxygen supply.

Steinschneider's findings prompted Richard Naeye, chairman of the department of pathology at the Hershey Medical Center of Pennsylvania State University, to look in SIDS victims for signs of underventilation—subtle markers that wouldn't be evident in a routine postmortem examination. In more than half the victims he found a minimal but significant increase in muscle tissue in the small pulmonary arteries, caused by the prolonged constriction in these arteries that occurs when the lungs contain insufficient oxygen. This tissue buildup in turn increases resistance to the flow of blood, creating a similar increase of muscle tissue in the overworked right ventricle of the heart.

These infants, Naeye proposed, were not responding as an adult would—with deeper breathing—to accumulations of carbon dioxide in the lungs. He examined the carotid body, a tiny organ in the neck that plays a role in restarting breathing when it has stopped, which is does for brief intervals during sleep. More than half the infants seemed to have an underdeveloped carotid body, possibly related to a structural problem in the brainstem. Such a defect could well develop before birth, resulting from some form of stress in the intrauterine environment.

The details of this theory, however, are virtually impossible to verify in live infants because the brainstem and carotid body are so difficult to examine. Another area that could play a role is the upper airway, the pharynx or larynx, where coordination of movement develops slowly in some infants. Obstruction of the air passage by the tongue and related muscles could be caused by certain toxins, for example, or by a deficiency of nerve cells in the portion of the brain that controls the tongue. If, in conjunction with this, the baby responds sluggishly to the air cutoff, the result could be fatal.

The only certain information about SIDS comes from infants who have already died. There are no known living patients to study and treat. This persistent obstacle makes steps toward an understanding of SIDS heartbreakingly slow. Researchers are currently trying to develop tests that will identify during the first few days of life the abnormality that leads to SIDS. When they can determine what the disorders are and correct them, Sudden Infant Death Syndrome will need a new name.

—Esther Mackintosh

How Does Anesthesia Shut Out Pain?

On October 16, 1846, Gilbert Abbott slept through his moment in medical history. He had entered the Massachusetts General Hospital in Boston to have a tumor removed from his neck, and the surgery attracted a crowd of incredulous physicians. Abbott was not given the proverbial bullet to bite, nor was he loaded with whiskey to dull his pain. He was simply told to inhale by one of the world's first anesthesiologists, a dentist named William T. G. Morton.

After the patient drifted into unconsciousness, "everyone present fully expected to hear a shriek of agony ring out as the knife struck down into the sensitive nerves," Morton's wife later reported. "But the stroke came with no accompanying cry." When the operation was over, the surgeon turned to his audience and said, "Gentlemen, this is no humbug." What he didn't say was how the anesthesia worked.

More than 140 years later, anesthesia remains an act of faith between the anesthetized and the anesthesiologist. Morton's demonstration that ether blocked the perception of pain ushered in a new era in surgery, and volumes have since been written on the proper use of general anesthetics. Thousands of these almost magical anesthetic compounds have been developed, and while the chemical structures of substances such as halothane, nitrous oxide (laughing gas), and barbiturates like sodium pentothal are as different as they can be, all cause the loss of consciousness.

Harvard pharmacologist Keith Miller explains that, metaphors to the contrary, sleep and the anesthetized state are not very similar. "When you're asleep, most of your reflexes remain intact, and you can still respond to many stimuli," he notes. "If you poke a mouse with a pin while he's sleeping, he'll wake up." An anesthetized mouse, however, will not. Neither will a patient.

"The primary effect is on nerve cells," says Miller, who has spent almost 20 years looking at what happens when anesthetic molecules meet neurons. When anesthesia is administered, muscles relax as the nerves controlling them are put out of commission. Perception is dulled, although the mind may be more receptive than previously suspected, disquieting news for glib-tongued doctors. The heart's output may drop slightly, and breathing is often labored enough to require assistance. In other cells, though, it's business as usual.

What's special about nerves? Researchers have known since the beginning of the century that an anesthetic's potency is directly related to how well it dissolves in olive oil, of all things. It turns out that anesthetic molecules have an affinity for fats, or lipids. Olive oil is a lipid, and a neuron's cell membrane is composed of two sheets of lipids, arranged in what is called a bilayer.

"Anesthetics disorder the bilayer and cause the membrane to expand," Miller explains, adding that when the membrane's expansion goes beyond what's known as a critical volume, the patient loses consciousness.

Researchers aren't sure whether the disorder is a cause or an effect of the anesthesia. There's certainly a connection. High pressure, they've found, can reorder a bilayer. If an anesthetized mouse is placed in a pressure chamber, it will soon come to its senses and act as if nothing ever happened.

Other changes in neurons also seem to play a role. Miller has been looking at what happens to a nerve cell's ability to transmit messages when an anesthetic is present. Neurons communicate via electrical and chemical signals. For news to pass from cell to cell, it must not only traverse the membranes electrically, but be ferried across the gaps, or synapses, between the nerves chemically. Chemical transmission may be almost instantaneous, but there is often a brief period right after firing during which the receiving cell is in a nontalking, or desensitized, state.

"We've recently discovered that anesthetics have one common property: they increase the number of desensitized cells," Miller says. In experiments with cells from electric fish—an animal often used in nervous system research—he found that anesthetic molecules interfere with the nerve's receptors for acetylcholine, a critical neurotransmitter that must flood the synaptic gulf between nerves and latch onto receptors across the synapse before messages can cross. The result is an increase in desensitized cells, causing a neurological low tide that won't begin to rise until the anesthetic is removed.

"We've learned that every anesthetic causing desensitization also disturbs the lipid bilayer," says Miller. But where these two observations fit in the nerve-to-brain-to-unconsciousness pathway is anybody's guess.

"We're still missing something basic," he observes. Like a detective probing around in the dark, Miller hopes to come across the "pharmacological fingerprints" leading to the essential mechanism orchestrating the molecular events. In the meantime, patients shouldn't lose any sleep over the matter.

—Bruce Fellman

Following Aspirin's Trail

Of all the drugs in a medicine cabinet, none is as familiar as aspirin. Most of us pop an aspirin as blithely as we take a shower or get a haircut. Americans swallow some 20 tons of the stuff everyday—enough to fill four good-sized dump trucks.

But even though aspirin has been around since 1899, only in the last few years have scientists begun to uncover how it works. Studies now show that aspirin may even help ward off heart attacks.

Aspirin seems to exert its major effects by blocking the production of an intriguing family of chemicals in the body called prostaglandins. The puzzle is to figure out what roles prostaglandins play in those symptoms for which aspirin is so effective.

Prostaglandins are hormonelike substances made by nearly every tissue in the body. They are essential to the smooth performance of many biochemical functions, including blood circulation, kidney function, digestion, reproduction, and possibly nerve transmission. Excesses of these chemicals, however, are involved in pain, inflammation, and fever—three symptoms for which aspirin has long been a remedy.

Prostaglandins aren't easy to examine. Normally, these elusive molecules occur in minute concentrations and decay in a matter of minutes. Because it is difficult to simulate in the lab what prostaglandins do in the body, much of what the biochemists know about them comes from observing what happens when people take aspirin or similar drugs.

Prostaglandins are found in suspiciously high levels in the inflamed joints of people with rheumatoid arthritis, a chronic disease with no known causes. Aspirin seems to combat the inflammation of arthritis by destroying the enzyme cyclooxygenase, which cells use to manufacture the prostaglandins. But prostaglandins alone may not cause inflammation. Unstable atoms known as free radicals—also present in damaged tissues—may be an accomplice. The role of the prostaglandins is shrouded in a network of biochemical interactions that scientists cannot yet trace.

The production of prostaglandins is also stepped up at the scene of an injury—whether a bruise, a cut, a burn, or even a headache. But it isn't clear whether prostaglandins cause the pain or simply sensitize the nerve endings that carry the pain message. In addition, prostaglandins may be acting on the central nervous system in some, as yet unknown, way.

Aspirin has long been favored to bring down a fever. Although the mechanisms of fever aren't completely understood, prostaglandin levels have been found to increase in the brain's temperature-regulating hypothalamus. Somehow, they seem to interact with neurons to raise the body's thermostat, initiating the fever. But no one can say exactly what the prostaglandins are doing. To complicate matters, some lab-induced fevers occur even after the suspected prostaglandins are blocked.

Whatever roles prostaglandins play, the discovery that aspirin blocks their synthesis has helped explain some of the drug's side effects. Aspirin takes a blanket approach to prostaglandins, wiping out all the cyclooxygenase it can reach in the body. Some prostaglandins, for example, protect the stomach lining from acid. By suppressing their synthesis, aspirin can cause anything from heartburn to ulcers.

Researchers are working on prostaglandin inhibitors that act more selectively, either on specific prostaglandins or in certain parts of the body. Already new drugs far outperform aspirin at relieving menstrual pain, which too is accompanied by an overproduction of prostaglandins; other drugs are more potent or more specific against diseases like arthritis.

The search for more sophisticated prostaglandin inhibitors also may yield drugs that are effective against much more than pain. Still, no matter what wonder drugs appear in the next few years, don't look for aspirin to disappear from the shelves. It's cheap, it's effective, and many people have come to be comfortable with it. But just how aspirin alleviates suffering remains an open question.

—*Steve Olson*

Why Do Cancer Cells Run Amok?

Normal cells are law-abiding. They reproduce only as often as necessary and then stop. Cancer cells, by contrast, are flagrant lawbreakers. They grow and divide indiscriminately—in the wrong place and the wrong time—heedless of all social constraints. What goes awry to transform one-time cooperative members of the biological community into perverse and disorderly outlaws, deadly disregarders of the rules essential to the well-being of the organism?

Normal human cells in a laboratory dish will inevitably divide only about 50 times and then never reproduce again—a phenomenon called the crisis of senescence or aging. Some cancer cells, however, will divide endlessly. The HeLa cell line has yet to show any hint of senescence, even after thousands of generations.

Speculation abounds—some of it wild-eyed and much of it based on sketchy or fragmentary evidence. There's little question, however, that something goes haywire in the genetic machinery to convert normal cells into biological desperadoes.

One guess is that an inactive cancer-causing gene dwells in a cell as part of the normal repertoire. When activated, this gene causes the cell to turn malignant. Viruses are known to do this, too. Some researchers are now hoping to identify such cancer-causing genes by matching the DNA, the stuff of genes, in malignant mammalian cells with the genes from known cancer-causing viruses. But even assuming that such a gene leads cells from the path of biological righteousness, what exactly does the gene do? Does it primarily affect another gene? Does some embryonic gene—one normally active only during the early stages of cellular growth—reactivate, causing unregulated and rapid cell division?

Or is the effect of the cancer-causing gene primarily on the cell surface? Does it somehow interfere with the cell's ability to respond to constraining signals from the external environment—by way of hormones, say, or direct cell-to-cell contact? Studies in laboratory cell cultures, for instance, suggest that some cancer cells may be limited in their ability to make fibronectin, a surface component thought to be fundamental to cell recognition and cell-to-cell binding. There's also evidence that certain cancer cells may have a hard time forming protein bridges, or gap junctions, with neighboring cells. These junctions, which provide hollow channels for the exchange of materials, i.e. information, between cells could conceivably serve as the major mechanism by which neighbors keep each other in check.

Others argue that cell transformation, rather than a direct consequence of cell-surface modification, may more likely be the result of a gene-evoked disturbance in the interior structure of the cell. This structure, the cytoskeleton, which consists of dynamic, cablelike assemblies that stretch from surface to nucleus, not only regulates cell shape but also shuffles surface proteins around. Electron microscopy has clearly revealed these strutlike structures to be far less organized in cancer cells than normal cells. How could such interior disarray promote a cell's normal-to-malignant conversion? The answer may lie, some suggest, in the uncoupling of the signaling system between surface and nucleus so that the genes never get the message to stop dividing.

Always there remains the ultimate mystery of first cause. Just what induces the cell to go astray? What kicks off the malignant process in the first place?

—Ben Patrusky

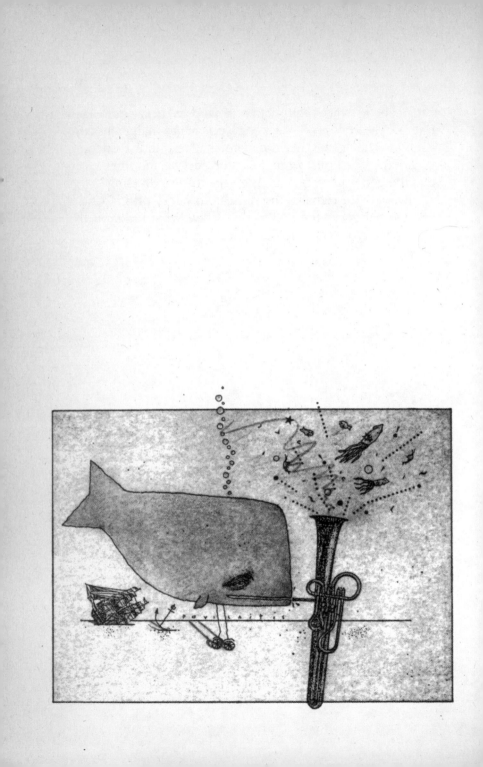

Part IV
FROM PLANTS TO ANIMALS

The Bird That Broke The Mold

The flights of Icarus, Orville Wright, and Sally Ride have been recorded for the ages. But nothing—no prehistoric bird watchers, no tracks in the sky—explains how the first bird came to fly. All that we have are clues from the fossil record, which invite a fundamental question: Did bird flight originate on the ground? Or in the trees?

In 1861 bones of the oldest known bird were first discovered in a limestone quarry in Bavaria. Five fossil specimens, each about the size of a crow, reveal a skeleton with feathers and wings, teeth, a lizardlike tail, and claws on the front edges of its wings. Named *Archaeopteryx* for "ancient wing," this archaic bird—not to be confused with earlier flying reptiles known as pterosaurs—dates back 150 million years and represents the "missing link" between reptiles and birds.

Discovery of the *Archeaopteryx* fossils stirred debate over whether birds are "glorified reptiles," as T. H. Huxley, the 19th-century English biologist, declared. This debate led to speculation, which has gone on for more than a hundred years, on how bird flight began. The fossil skeletons provide evidence that *Archaeopteryx* could fly, but whether it lived on the ground or in the trees, or even if it was the first bird, is a mystery.

Among *Archaeopteryx* watchers, a consensus favors the trees-down, or arboreal, theory for the origin of bird flight. Walter Bock at Columbia University has long contended that the ancestors of *Archaeopteryx* were tree-dwellers. According to his theory, leaping between branches eventually yielded flight that started as gliding and later became flapping.

But aerodynamics don't support gliding as a precursor to bird flight, according to the calculations of Gerald Caple, Russell Balda, and William Willis. This trio from Northern Arizona University finds no aerodynamic reason for a species of gliders—flying squirrels, for example—to change. "A glider does worse by trying to flap, than by not flapping at all, if it's trying to extend its flight," argues Balda.

Instead of envisioning flight as the culmination of leaping and gliding, Balda and his colleagues support a ground-to-trees, or cursorial, theory. This theory holds that the forelimbs of ground-dwelling ancestors of *Archaeopteryx* gradually evolved into incipient wings for leaping up to catch insects. Eventually, supporters argue, the proto-wings of these animals became capable of sustained flight.

Aerodynamics notwithstanding, the laws of gravity would seem to support the arboreal theory. Its proponents claim that it makes more sense for flight to originate in trees with the assistance of gravity, rather than on the ground against gravity. Indeed, Jeremy Rayner at the University of Bristol objects to the cursorial theory on the basis that a ground-dwelling insectivore would need an improbably high running speed to take off. And other supporters point out that all present-day vertebrate flyers, with the exception of flying fish, are arboreal.

The leading champion of the ground-dwelling theory, John Ostrom of Yale University, relies on anatomical evidence to support his theory. He describes the arboreal theory of Walter Bock as "a very elegant model and perfectly logical." Yet he believes that the well-developed hind legs of *Archaeopteryx* show specialization for ground dwelling. Supporters of the arboreal theory, on the other hand, argue that the claws on the wings of *Archaeopteryx* closely resemble the claws of squirrels and other tree-climbing animals. Ostrom concedes the point but maintains that *Archaeopteryx* also appears to resemble a type of dinosaur, known as a coelurosaur, which was a ground dweller. Were it not for its feathers, says Ostrom, *Archaeopteryx* could be classified as a ground-dwelling reptile.

Feathers, and the reason that they evolved from reptilian scales, are another source of contention. Did they evolve for flight? Or for insulation? According to the arboreal theory, the tree-dwelling ancestors of *Archaeopteryx* developed feathers to extend the distance of leaps between branches. But some scientists hold that feathers developed in the extremities, where heat loss is greatest, for thermal insulation.

Bock remains immovable. "As far as I'm concerned, the terrestrial theory has been grounded, if not buried." More than a hundred years of speculation, however, isn't likely to end here. Given the fossil record we have, Ostrom says, "It is highly improbable that we will ever solve the mystery."

—*David Savold*

How Do Animals Hibernate?

Human beings go into suspended animation only within the pages of science fiction, yet hibernating animals manage this near deathlike state for four to six months each year. When a ground squirrel, for instance, curls up for the winter, its body temperature drops from the normal mammalian level of about 98 degrees Fahrenheit to as low as 34 or 35 degrees, and its heart, over a period of three to four hours, slows from 350 beats a minute to two to four beats a minute. How do animals survive this drop in body temperature and metabolic level? And how do animals hibernate in the first place?

"We don't really know," says Roland Aloia of the anesthesiology departments of Loma Linda University School of Medicine and Pettis Memorial Veterans Hospital in Loma Linda, California. Aloia hopes to answer these questions by studying biochemical changes in the cell membranes of hibernating ground squirrels.

Hibernation conserves energy. It is a way animals can avoid the lean, cold winter. As Aloia points out, "An animal that can reduce its metabolic activity by 98 percent when there isn't much food around is way ahead."

Although people associate hibernation with bears, only small mammals like ground squirrels, hamsters, hedgehogs, woodchucks, and bats can enter the torpor that defines true hibernation. Some animals, including various species of snakes, land snails, and lizards, enter a state of dormancy, or estivation, in the summer when water is scarce. In bears, the ratio of surface area to body mass is not great enough to dissipate the heat required to lower the animal's temperature to hibernation levels. A wintering bear's temperature rarely falls below 86 degrees Fahrenheit, so they are rather easily awakened; it takes a true hibernator several hours to arouse from a near-freezing body temperature.

For years scientists thought that the cooler weather of autumn triggered the onset of hibernation, but goldenmantled ground squirrels kept in a constant temperature of 95 degrees Fahrenheit show signs of hibernation every 300 to 360 days or so. The squirrel cannot lower its body temperature in such heat, but it will lose weight as if in hibernation.

This seems to indicate the presence of an internal body rhythm, according to Eric Pengelley at the University of California at Riverside, who has found that many other ground squirrel species also have self-contained clocks. The internal clock, which may or may not exist in other hibernators, apparently directs the metabolic changes that occur in ground squirrels every year.

Researchers have spent a great deal of time in the last two decades trying to find what triggers hibernation. At the Stritch School of Medicine of Loyola University in Maywood, Illinois, Albert Dawe and Wilma Spurrier isolated a substance from the blood serum of hibernating woodchucks and one ground squirrel species. When Dawe and Spurrier's "trigger substance" is injected into ground squirrels, it results in hibernation.

Trigger substance may work in combination with another secretion called antitrigger. Dawe and Spurrier hypothesize that antitrigger in the bloodstream keeps an animal active throughout spring and summer. The production of trigger substance continues, however, and by fall it overrides antitrigger, and the animal hibernates. The animal becomes active again when enough antitrigger is produced in the spring. As Dawe and Spurrier see it, these substances hold the key to understanding the complex mechanisms of hibernation.

Henry Swan at Colorado State University has extracted a hormone he calls antabalone (for antimetabolic hormone) from the brains of squirrels and other animals. When antabalone is injected into rats, which do not hibernate, it lowers their metabolic rate: this in turn lowers their body temperature about 42 degrees Fahrenheit.

To prepare for hibernation, a ground squirrel more than doubles its body weight and prepares a burrow about three feet deep—a depth that allows for a nearly constant temperature regardless of surface variations. Then it digs a small side pocket, or hibernaculum, barely larger than itself, where it will spend the winter curled up in a ball with head and paws tucked underneath its body. The tight pocket helps the hibernator conserve heat, but it also severely reduces the dissipation of carbon dioxide. Even though a hibernating squirrel may breathe only once every several minutes, its system remains well oxygenated and shows no evidence of dangerous carbon dioxide buildup—how, no one knows.

The hibernator's periodic arousals also puzzle researchers. At varying intervals—about every 14 days in ground squirrels—the animal's body temperature returns to normal. It may or may not eat or defecate but always urinates: after 12 to 24 hours of activity, the squirrel returns to hibernate for another 14 days. During these active times, the ground squirrel uses up 80 to 90 percent of the total calories consumed during hibernation.

Roland Aloia has been looking for answers to some of the mysteries of hibernation by studying biochemical changes in the membranes of cells that make up different organs in hibernating animals. In membranes, which control the movement of substances in and out of cells, fatty substances called phospholipids undergo structural change so as to better function at low temperatures. Many of the fatty acids attached to the phospholipids become less saturated, and this may improve the conduction of electrical impulses in the brain and heart at hibernation temperatures.

"Hibernation itself is remarkable, but the most notable thing, in my opinion," says Aloia, "is the ability of the membranes to remain in excitable states at low temperatures. For instance, our hearts would simply stop beating at about 59 to 68 degrees Fahrenheit, but the hibernator's heart continues to beat at 34 degrees. And the hibernator's nerves continue to conduct impulses at that low temperature. How does the squirrel manage? It would be nice to know."

—Carrol B. Fleming

How Do Plants Know
Which End Is Up?

A seed falls, and finding the weather hospitable, it swells, splits its protective coat, and sends a shoot and a root out into the air. Suddenly and without fanfare, the shoot curves upward and heads for the sky, while the root turns downward and angles its way into the soil.

This uncanny ability to sense gravity and act accordingly, called gravitropism, has long fascinated botanists. But so far no one has been able to explain how plants get their directions straight. Darwin tried rigging shoots and roots with tiny pens, then watched as they traced an elegant ballet denoting differences in cellular growth rates. He saw that in the shoot, the lower level cells grow faster than the upper level cells. But in the roots, growth is more rapid in the upper level cells.

The cause of these plant curvatures eluded Darwin, yet he had one tantalizing lead. Roots, he found, became confused if their caps, the specialized helmets that offer protection from the hazards of digging, were removed. Was there a "brain" in these hard hats that somehow directed the show? And how did shoots, which don't wear caps, fit into the orientation picture?

The mystery appeared to have been solved in the 1920s with the discovery of auxin, a hormone that regulates plant growth. Auxin has a split personality—it stimulates growth in shoots but inhibits it in roots, except where concentrations are lowest. If gravity somehow caused the hormone to move downward, auxin would pool up in the lower-level cells of each plant part. This change in hormone distribution could cause the change in directions.

But botanists now disagree over how much, if any, of an auxin pooling occurs before curvature. Instead, the mineral calcium seems to play a more important role. Plant cells lengthen during growth, and while auxin softens the cell walls, calcium strengthens them. Too much calcium, however, inhibits growth by making cell walls stiff and rigid. During bending, high calcium levels are found in the upper portion of the shoot as well as in the lower portion of the root cap. If a growing root or shoot is placed horizontally, the action resembles a boat powered by one oar. Instead of making headway, each plant part turns.

Michael Evans of Ohio State University is looking at how this excess calcium is involved in a plant's response to gravity. The root cap, he explains, is rich in cells that contain amyloplasts, heavy capsules filled with starch and large amounts of calcium. These capsules respond to the Earth's relentless tug by dropping like stones to the cell floor. There, Evans suspects, amyloplasts dump enough of their mineral load to turn on special chemical pumps that move excess calcium to where it's needed—the lower cells of the cap.

"If you prevent calcium's movement," he says, "there'll be no response to gravity. And if you simply put calcium on one side of the cap, where it is quickly absorbed, the root bends toward that side just as if gravity were being exerted."

Shoots, of course, don't have caps but they do have amyloplasts. When they descend, calcium travels to the *upper level* cells.

How does the mineral know where to pile up? Stanley Roux at the University of Texas notes that differences in electrical charges between the tops and bottoms of cells seem to polarize the cells. If enough of these living batteries were lined up in a row, their combined charge might be strong enough to attract calcium atoms or hormone molecules with the opposite charge.

"It's all speculation," says Roux, "but it's tempting to think that, because of this polarity, a plant *always* knows which end is up. This much is true—if you put plants in an electrical field, you can get them to grow in any direction."

But if orientation does depend on electrical currents, we still don't know what force aligns them. Curvature becomes visible within 10 minutes, and so far botanical sleuths have traced a trail of events back to the one-minute mark—when amyloplasts drop. There, the trail ends.

—*Bruce Fellman*

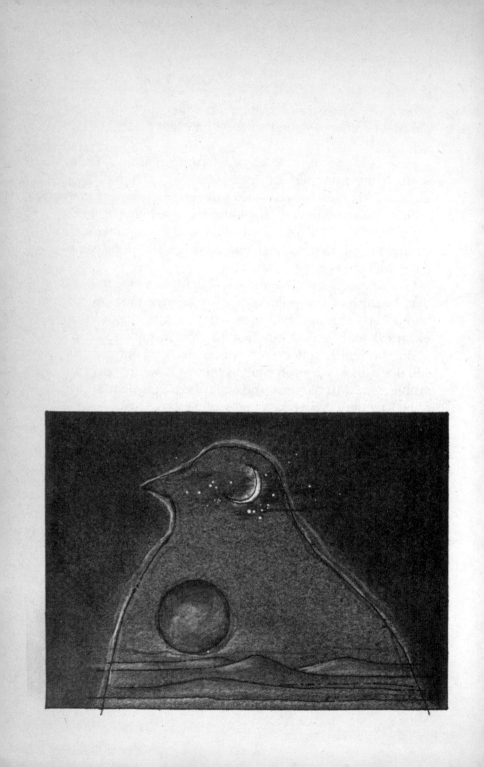

How Do Birds Find Where They're Going?

The melancholy appearance of geese passing south under low autumn skies is as much a mark of the turning seasons as the first robin of spring. Some of us pay more attention to these things than others, but few are more drawn to the seasonal movement of birds than ornithologists who have for years been attempting to understand one of migration's most vexing riddles: How do birds know which way to go?

Billions of the world's birds migrate each year, some on journeys of thousands of miles. These restless voyagers have been strapped with radio signaling devices, shadowed by airplanes, and banded by the millions. With banding, humans can even keep pace with such energetic migrants as the arctic tern—the long-distance champion at 22,000 miles a year. Yet there is no sure way to tell what sights, smells, sounds, or other signals guide a bird to the same wintering place year after year.

Ornithologists take different routes in exploring migration. Perhaps the simplest theory is piloting, that birds navigate using landmarks such as rivers and mountain ranges, and only piece the trip together as they go—what some call steeplechasing. Piloting, however, can't account for the results of experiments like the one done in the 1950s involving a sea bird called the Manx shearwater. Plucked from its nest on an island off the coast of Wales, it was transported to Boston, Massachusetts, and released. In 12½ days it had returned to its nest after crossing 3,050 miles of featureless Atlantic.

Most researchers agree that birds take directional cues from the environment over which they fly. Besides the sun and stars, birds may rely on changes in barometric pressure, wind direction, odors; even low-frequency sounds created by wind, thunderstorms, or distant ocean surf. And scientists have observed that different birds seem to use cues in different order. While it makes sense that birds take advantage of these aids, no one can explain how birds are programmed to do so or to what extent their ability is learned or innate.

Although most migration is carried out at night, a popular cue among daytime fliers is the sun, which birds use as a compass. By eyeing the sun's position and combining it with their own circadian rhythms, or internal clocks, they find and maintain the correct heading—similar to the way a hiker in the woods knows that the sun on his left in morning means he's moving south. Tests on pigeons indicate that the sun affects the direction they take. But when the sun is obscured pigeons take their heading from another cue, probably Earth's magnetic field. When test birds deprived of the sun were equipped with bar magnets intended to disrupt their readings, they indeed became disoriented.

Tony Lednor of Cornell University points out that there are several theories on how birds might detect the Earth's magnetic field. But the most exciting is linked to a discovery in 1979 of magnetite, a magnetic material, in the heads of pigeons. Magnetite, notes Lednor's colleague Charles Wal-cott, may be part of a sense organ that gives birds their orienting skills. A number of experiments are underway to determine just how the magnetite functions, but it isn't yet known which birds possess it or how important it is to navigation.

Most experiments are hampered by the fact that migrants seem to use more than one way of getting around. "For a long time," says Lednor, "research tended to look at bird navigation as having a single cause. What we're studying now is how they use a whole network of cues." A bird deprived of a preferred orientation source, such as the sun or stars, may switch over to another cue in the same way airplane pilots rely on instruments in fog. All of which has made the ornithologist's job more mind reader than fact finder. Despite the certainty of their yearly sweeps across the world, the interaction between migrating birds and the cues that guide them is still elusive.

—Patrick Cooke

Monarch Mystique

How does a monarch butterfly migrate 2,000 miles or more to a winter refuge it has never seen, the same refuge its great-grandparents left the prior spring? "It's amazing," says University of Florida lepidopterist Lincoln Brower. "This is just something genetically programmed into the butterfly."

In the heat of summer in the eastern United States and Canada, monarchs hatch, mature, mate, and die in less than two months. The cycle repeats three or four times in the season, leaving the last—and only surviving—generation to find its way south in winter ahead of the killing cold. At the end of the journey, they converge in the mountains of central Mexico at a handful of sites, some of which measure not more than an acre. While monarchs home faithfully year after year to these dots on the map, they have no known means of finding their way. Their progenitors that came north in spring have long since died.

It wasn't until 1975 that scientists even knew where eastern populations of the butterfly wintered. That year, Fred Urquhart of the University of Toronto, after four decades spent chasing monarchs, found millions of them wintering at a 10-acre site 10,000 feet up in the mountains of

Mexico. He recalls being "dumbfounded" at the sight of the forest so laden with monarchs that tree branches bent with their weight. Since then, 15 other sites, ranging from one to 10 acres, have been found in a 100-mile radius.

At the heart of monarch migration, it appears, lies the insect's response to daylight. When the days lengthen in spring, sex hormones flow, and the monarchs mate. The shortened days of August then halt that flow and quicken the butterfly's lust to wander.

Migrating monarchs fly only during the day, and two of Brower's colleagues at the University of Florida have discovered a possible means of navigation. Bruce MacFadden and Douglas Jones detected magnetic material, iron oxide grains called magnetite, in the bodies of monarchs. This, they suspect, serves the butterfly as an internal compass by which it can orient to the Earth's magnetic field.

But while daylight and magnetite may send the monarch in the right general direction, these factors do not explain how successive generations of butterflies return year after year to the same small, select sites.

The best clue as to why this occurs, it seems, is bound up in the insect's natural history. Ranging from Canada to Argentina, the monarch is a North American pioneer; it is the only species in a family of tropical butterflies to gain a foothold in the temperate climate of the northern United States and southern Canada. These areas provide a rich supply of milkweed, a broadly diversified plant upon which the adult monarchs depend for nectar. In addition, after their dormant winter in Mexico, monarchs deposit their eggs on the sprouting milkweeds that line the path of their return trip each spring. The caterpillars that emerge are then nourished by the plant's leaves as they develop into adults.

But during the winter, the Mexican sites form a niche for the species, giving it a tenuous yet vital lease on life in the midst of the forbidding cold. At 10,000 feet in the mountains of Mexico, the January nights can grow cold enough to freeze several million monarchs. But the insects gather in forest so dense, says William Calvert, Brower's colleague, "that the trees touch to form a blanket." If this forest is even selectively logged, it loses its insulating properties and dooms the butterfly.

Thus, high in the mountains of Mexico, there is an ecological space that the butterfly exploits. It is, however, a small, specific niche, and apparently the only place where the monarch can survive the winter.

The urge to find that place therefore becomes one and the same with the urge to survive. Somehow, that urge in all its power passes from generation to generation, enabling the monarch each winter to find its way south to those few places that will shelter and nurture the species for yet another year. They are places the butterfly has never seen but knows it must find.

—Edward O. Welles, Jr.

How Do Whales Catch Their Dinner?

Not since Jonah took his fateful plunge into the mouth of "a great fish" has there been anyone else to provide cetologists with a firsthand account of how whales catch their prey. It is a mystery as old as the Old Testament and as elusive as the marine mammals themselves.

Whales are categorized according to whether they have teeth (odontocetes) or baleen (mysticetes), a comb-like structure attached to their upper jaw. Baleen whales, it is known, feed themselves by gulping or taking in water and using their baleen as a strainer to retain the fish in the water. But how some 67 species of toothed whales catch their dinner confounds cetologists. Sperm whales, for example, feed primarily on squid. But squid removed from the stomachs of these 15- to 45-ton mammals show no teeth marks. How does this ponderous animal manage to capture its fast-moving prey?

Cetologists are also puzzled by the feeding habits of the male narwhal. He somehow manages to swallow polar cod and Greenland halibut despite an eight- to ten-foot tusk that protrudes straight out from his upper lip. But the real now-you-see-him-now-you-don't gourmet of the odontocetes is the male strap-toothed whale. This woeful leviathan is hard-pressed to even open its mouth because of a pair of lower teeth that curve up and over its upper jaw.

Attempts to explain how toothed whales catch their prey aren't new. In the thirteenth century, the medieval monk Bartholomaeus Anglicus wrote in his famous encyclopedia, *On the Properties of Things,* that a whale produces an amber-like smell to lure fish into its mouth. Whalers in the nineteenth century believed that prey were attracted to the brightness of the whale's mouth. But within the last two decades, a new hypothesis has been gaining attention. Many cetologists now believe that the toothed whales stun their prey with strong bursts of sound.

In 1946, it was discovered that porpoises, like bats and submarines, use sound or echolocation to locate objects. Porpoises, dolphins, and other toothed whales emit sounds and detect a potential victim from the resulting echoes. It has been proposed that some sounds reach intensities powerful enough to debilitate prey.

Kenneth S. Norris, from the Center for Coastal Marine Studies at the University of California in Santa Cruz, has done studies that support this hypothesis. Evidence indicates that dolphins have evolved the ability to use sound as a weapon. Modern species of dolphins emit a focused beam of sound, as would be necessary to hit a target, compared to the broad pulse of sound emitted by archaic species. (Archaic whales are living species that show little evolutionary change.) Norris also notes that many modern species of toothed whales have fewer and less specialized teeth, unlike the rows of sharp teeth found in ancient whales, which might have been used for snaring prey. His theory is that sound has replaced teeth as a hunting weapon. Fishing with explosives, he suggests, is an apt comparison of how toothed whales use sound. "The only way we could mimic the strength of dolphin sounds," Norris says of a past experiment, "was to use blasting caps." The pressure of the sound waves might even kill some prey.

How whales produce sounds, however, adds another layer to the mystery. Whales have no vocal cords. The clicking sounds produced for echolocation are thought to resonate from the forehead, but no one is sure how. And toothed whales produce a variety of other sounds described as whistles, chirps, moans, barks, yelps, yaps, squeaks, and squeals.

Measuring whale sounds hasn't been very helpful in proving the hypothesis. Because of difficulties in underwater acoustics, no one has recorded sounds of intensities high enough to stun prey in the ocean. James Mead, a cetologist at the Smithsonian Institution, stresses that sound debilitation of prey by toothed whales "is a new theory and not a demonstrated fact. All we've got is speculation." In addition, the sound-stunning strategy may not hold true for all odontocetes.

The sonar capabilities of toothed whales remain a mystery largely because of the environment in which whales live. Most toothed whales feed in deep water and can rarely be observed for any extended period. Norris and his colleagues are trying to record whales as they feed along the Pacific coast. Until experiments like this provide new clues, the mystery will stay in the oceans.

—*David Savold*

Why Do Whales Run Aground?

As American whale experts gathered for a conference in the late 1970s in Corvallis, Oregon, a pod of 41 sperm whales was slowly dying on a sandy beach an hour away. The scientists stayed up all night plotting ways to haul the animals back out to sea. After rushing to the site at daybreak, they realized their task was hopeless. Available helicopters weren't strong enough to pull even the smallest whale, which was 30 feet long and weighed 15 tons. The shallow beach prevented tugboats from getting near. And no one was positive that, if rescued, the animals would stay at sea. Freed whales often turn around and swim determinedly back to shore.

By noon that cloudy day, some 5,000 people had gathered on the beach to watch the silent ebb of life. "It was horribly frustrating to see them die," says Bruce Mate, a marine biologist at Oregon State University. "The only way to make their deaths meaningful was by doing research on the carcasses to learn more about why whales strand."

Why herds of seemingly healthy whales run aground is as puzzling today as it was 2,000 years ago, when Aristotle wrote that whale strandings happen "without apparent reason." Of 78 species of whales, only 11—all toothed whales, including dolphins—strand in groups of three or more animals. Though about five mass strandings are reported around the world every year, the incident in Oregon remains the best documented because scientists were on the scene within hours. They took blood tests and samples of teeth, skulls, and other body parts from 10 whales before the rest decomposed. But their research has only weakened most theories on strandings. "With just a little data, you can erect nice hypotheses that seem to work," says James Mead, a cetologist at the Smithsonian Institution. "Now we have enough information to know that most of our theories don't fit."

The explanations vary from pathological—that roundworms nesting in whales' ears impair their navigation (though similar infestations occur in cetaceans that don't mass strand)—to mystical—that the whales are committing suicide. Some say that whales are following primordial routes of navigation that steer them into coasts that didn't exist 50 million years ago. Another theory is that whales are reverting to the behavior of their amphibious ancestors, who probably sought shelter on land when they were sick, injured, or attacked by predators. But Mate, who led the Oregon research, calls these hypotheses absurd because species with such self-destructive instincts would have died off long ago.

A leading hypothesis—one Mate doesn't shrug off—is that whales strand because they are social animals with an instinct to herd. If the leader of a pod beaches because it is sick or swims too close to shore, the others will follow. Unfortunately, many strandings, as in Oregon, occur on isolated beaches far from human eyes that can determine which animal beached first. Nor is it clear if there is a single leader. The pod could be split according to family ties.

Several unusual circumstances may be partially responsible for the Oregon strandings, says Mate. Plankton blooms in the water that day were so dense that even ships with sophisticated radar couldn't locate the ocean floor. Also, a strong upwelling of cold water from the ocean bottom, which normally runs parallel to the shore, took a sharp bend straight into the beach. The nutrient-rich water may have enticed the feeding whales to their death.

Although everyone agrees that beachings occur under different circumstances and for a variety of reasons, no one is sure what makes the lethal combination. "The current explanations are all less than likely," says Smithsonian's Mead, "but we're faced with a less than likely situation."

—Blythe Hamer

Part V
THE PLANET EARTH

Why Is The Earth Neither Too Hot Nor Too Cold?

The climate of the Earth has varied considerably over time. One hundred million years ago, ichthyosaurs were frolicking in the warm, shallow seas of what is now Kansas. And Ohio was once a frozen tundra where huge woolly mammoths trod. Yet what puzzles many scientists is not that the climate varies but that it has remained as stable as it has.

The Earth, they point out, is quite literally poised between fire and ice. Consider, for example, what would happen if we somehow moved the Earth slightly closer to the sun.

As the oceans grew warmer, more and more water vapor would begin to steam into the atmosphere. Once there, the vapor would begin to act like glass in a greenhouse, preventing heat from radiating back into space. So the Earth would grow warmer still, until oceans began to dry and the carbonate rocks—the limestones and dolomites—began to bake in the heat and release carbon dioxide.

The greenhouse effect caused by this gas is famous: By burning fossil fuels we are already releasing enough carbon dioxide to warm the climate measurably. The carbonate rocks, however, contain billions upon billions of tons of it, enough to trigger a "runaway greenhouse." In the end our planet would become a twin of unfortunate Venus, the next planet inward to the sun: a gaseous, dry, searing hell, its surface covered with clouds, oppressed by a massive atmosphere of carbon dioxide, and hot enough to melt lead.

Suppose, on the other hand, we moved the Earth further out from the sun. As the planet grew colder, glaciers would grind southward over Canada, Europe, and Siberia, while sea ice crept northward from Antarctica. The ice would reflect more sunlight back into space, cooling the planet even more. Step by step, the ice would extend toward the equator. In the end, the Earth would gleam brilliantly—but its oceans would be frozen solid.

Thus, the climate is balanced precariously indeed—so precariously that many geologists now believe that tiny, cyclic variations in the Earth's orbit, known as the Milankovitch cycles, were enough to have triggered the ice ages.

But geologists also have found fossilized marine microbes in rocks more than 3.5 billion years old, and they assure us that the oceans of the Earth have remained warm and liquid throughout its 4.6-billion-year history.

Perhaps this is a lucky accident—after all, if the Earth had not formed at just the right distance from the sun to have liquid oceans, we would not be here to worry about it. But the astrophysicists point out that things aren't quite that simple.

The sun, they say, is a quiet and stable star. But like others of its type, it is inexorably getting hotter with age. In fact, it is about 40 percent brighter now than when the Earth was born. So how could the climate possibly stay constant? If the Earth is comfortable now, then billions of years ago, under a colder sun, the oceans must have been frozen solid. But they were not. On the other hand, if the oceans were liquid then, why has the sun not broiled us into a second Venus by now?

One theory, advocated by a number of biologists, is that the early Earth started out with a good deal more carbon dioxide in the air than it has now, which gave enough of a greenhouse effect to keep the planet warm even when the sun was cool. If nothing had changed, the greenhouse eventually would have "run away" as it did on Venus. (There is some evidence that Venus started out with oceans much like ours.) But fortunately for us, about two billion years ago, certain blue-green algae devised a way of taking carbon dioxide out of the air and turning it into organic carbon compounds. We call that process photosynthesis. In the eons since, as the algae and their descendants, the plants, have evolved and multiplied, the decreasing levels of atmospheric carbon dioxide have just about kept pace with the warming sun. Thus, say the biologists, it was life that saved the world for life.

This idea is far from proven, of course. And lovely as it is, in a way it would be rather sad if it were true. For if the continuation of life on a given planet depends on an evolutionary accident, followed by a remarkable fine-tuning of its atmosphere to a warming sun, then the hopes for finding other intelligence in the universe must be slim indeed. Or perhaps not—for the fact is that no one really knows why the climate of the Earth is stable.

—*M. Mitchell Waldrop*

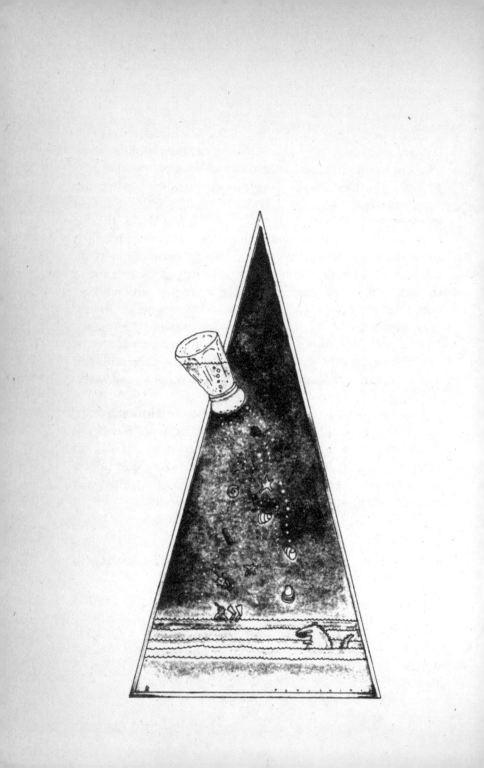

Why Is The Sea Constant?

Two hundred million years ago the seven seas were one, surrounding the vast continent of Pangaea. Fifteen thousand years ago the ocean was 400 feet lower, its missing water locked up in bloated polar ice caps.

Despite such changes, the sea has always kept its chemical poise. A triceratops blundering into the ocean would have tasted the same saltiness that startled children do today. A mass spectrometer magically transported back in time would have measured much the same elements. In fact, by studying sedimentary rocks, oceanographers conclude that the composition of the ocean hasn't changed much over the past 700 million years.

Why has the makeup of the ocean stayed so constant? A multitude of processes, including biological activity, affect the sea's chemistry. Yet the relative quantities of the elements found in seawater generally have remained the same. Understanding why the sea's composition remains stable could yield clues not only to the history of the Earth but also to the origin of life.

All the Earth's elements can be found in the ocean. Most of the material dissolved in seawater is regular table salt, sodium chloride. The other common elements that form salts—sulfate, magnesium, calcium, potassium—come next in decreasing order, followed by the remaining known elements.

Life, which probably began in the sea, remains a chemical choreography of many of the sea's elements. Even though 350 million years have passed since our ancestors crawled onto land, the salt content of human blood bears a haunting resemblance to that of the ocean. The calcium that goes into making the shells of marine organisms also strengthens the skeletons of land dwellers. And hydrogen, oxygen, carbon, nitrogen, and phosphorus make up DNA, the chemical informant for all life.

Many of the elements in seawater are poisonous to living systems. Lead, for example, disrupts the membranes around cells, destroying their ability to protect the cells from invasion.

Did the first traces of life simply use the elements that they found around them, like inventive cooks in someone else's well-stocked kitchen? Or did early organisms actively change and stabilize their watery habitat, turning the ocean into a giant incubator to nurture life in its infancy?

The answers to such questions are lost in the jumble of geologic time. The grindstones of plate tectonics have chewed up much of the evidence, and the crumbling sediments that geologists do dig up reveal little. But early life clearly had an important effect on its surroundings. For instance, the first oceanic plants completely remade the atmosphere. By releasing oxygen during photosynthesis, they created an Earth that could sustain respiration, paving the way for the evolutionary pageant to follow. This biological milestone must have transformed the ocean, but no one knows precisely how.

Living things also affect the present makeup of the ocean. Most of the calcium and silicon dissolved in seawater leaves it as the shells of microscopic organisms. And plankton floating through surface waters scoop up the nutrients nitrogen and phosphorus. When the plankton die, bacteria release the nutrients into the deep. Wherever this deeper water wells toward the surface, as off the coast of Peru, life in the ocean prospers.

Many reactions in the ocean do not directly involve living things. When river-borne minerals mix with seawater, for instance, they are chemically transformed into ocean floor sediments. And chemical activity at the extremely hot vents found along mid-ocean ridges takes some elements out of seawater and puts others in. But in the cumulative makeup of the ocean, life has its role: It acts as a central point through which many chemical cycles pass, joining these cycles into an interdependent system. This could help stabilize the composition of the ocean, though we do not know for sure.

There may be a risk in our ignorance. If we continue to increase the amount of carbon dioxide in the atmosphere by burning fossil fuels, the cycling of carbon through the ocean could conceivably change, affecting the balance of marine life.

So far, however, the chemistry of the sea has remained constant. But no one has completely fathomed why.

—Steve Olson

What Makes The Field Flip?

The Earth's magnetic field is fading. If the decline in strength continues at its present rate, in 1,200 years the world's compasses will become confused. For a time they will point anywhere and everywhere. Then slowly, after decades or centuries, they will begin to straighten out again—but pointing south instead of north.

The result would be the next great magnetic field reversal. It has happened before, many times. Geologists have found the evidence in magnetized rocks millions of years old. The phenomenon is clearly telling them something profound about the Earth's inaccessible core. The question is, what?

The core of the Earth is made of iron and nickel, most of it molten. This metallic fluid is in constant motion, and the motion somehow sets up electric currents that generate the Earth's field. The details of this molten action and the flips it causes in the magnetic field are not known.

Some clues to what is happening come from the external structure of the field, which is immense. It surrounds the Earth and extends hundreds of thousands of miles into space. The magnetic field can be thought of as a series of imaginary or invisible lines arcing through space from the south magnetic pole in Antarctica to the north magnetic pole in Canada, then looping through the core of the Earth to emerge again in the south.

In practice, the Earth's field is lumpy and skewed. The magnetic poles lie 11 degrees of latitude from the true poles, and there are twists and eddies in the field, regions where a compass can point as many as 20 degrees from true north. Mariners have mapped these regions since the 1600s, lest their compasses lead them astray. From their records we know that the magnetic field fluctuates greatly in strength and drifts westward about 12.5 miles per year. To modern scientists this suggests that the molten fluid in the core is moving at the rate of about one-half a millimeter a second—half the length of a football field each day.

On a cruder scale, geophysicists have traced the field back 30 million to 50 million years by studying the magnetism frozen into ancient lavas: Iron atoms in the molten rocks tend to line up with the prevailing magnetic field as the material solidifies. The record shows reversals occurring at random, at intervals of from 30,000 to one million years. The field simply turns inside out, fading away over a period of about 10,000 years and then starting to grow in the other direction.

Many geophysicists probing the cause of reversals now believe that the magnetic field we observe on the Earth's surface—so feeble that a toy horseshoe magnet is easily 100 times stronger—is really just a sideshow. Most of the magnetic action goes on in the Earth's nickel-iron core. According to the most popular explanation, the dynamo theory, the part of the field that runs through the core is trapped in the molten, electrically charged fluid and dragged along by the Earth's rotation. As a result it doesn't just pass straight through the core; it wraps around and around, stretched tight like a bundle of rubber bands into immensely strong hoops of flux.

The idea is that the convection of molten metal welling upward from the core of the Earth pushes tiny loops of the tightly wound magnetic material to the surface, where they arc out into space to form the familiar field and then dive back into the core to be wound up tightly again. Thus the field maintains itself.

As for what might cause the flip, one version of the theory is that it may be the random nature of the convection: If a few more loops crop up in one place than another, the tiny bits of field they push to the surface would start to loop in the opposite direction. Another possibility is that the reversals are not random and would be quite predictable if we knew enough. Perhaps the electromagnetic couplings of the boiling fluid within the Earth are so complex that the reversals just look random. If so, perhaps one day scientists will be able to tell us when the next great reversal is due.

But for now all we can do is watch our compasses and wonder what is going on in the great molten heart of the Earth.

—*M. Mitchell Waldrop*

Mother Earth's Uncertain Future

And when the thousand years are ended, Satan shall be loosed out of his prison.

—Revelations 20:7

Science, too, has its apocalyptic prophecies, though we'll probably have to wait billions of years rather than thousands to see them fulfilled. Scientists agree that at some point in the very distant future, plants and animals will no longer be able to survive on Earth. But predicting when that will be is impossible.

The sun's life-span sets the upper time limit on Earth's habitability. At the moment, the sun is living through a comfortable middle age. For more than four and a half billion years, it has been supplying the Earth with heat and light, as it should continue to do for about the next five and a half billion.

After that, when the sun is roughly 10 billion years old, it will exact a fearsome toll on its nearby dependents. Its interior depleted of hydrogen, a star's most abundant fuel, the sun will metamorphose in a last-ditch attempt to burn other fuels. It will shine 400 time brighter than it does now and swell to 100 times its size—from Earth it will seem to fill the sky. Its color will deepen to blood red. Mercury, the innermost planet, will be enveloped by the expanding ball of fire. Though charred to cinders, Venus and Earth will survive, but any remnants of life will be cremated.

It is highly unlikely, however, that any of our descendants will be around to perish in Earth's fiery finish. Long before, all plants and animals will probably succumb to a less violent threat. Throughout its history the sun has been getting brighter, and in the future the Earth will become so hot the oceans will begin to boil, mantling the planet with a thick blanket of steam. Entirely new kinds of organisms may evolve to live in the cooler upper fringes of the atmosphere. But the temperature at the Earth's surface, over 800 degrees Fahrenheit, will be far too hot for any known type of organism to survive.

Climatologists disagree on when the Earth will be transformed into a giant steam bath. Some say it won't happen until close to the end of the sun's lifetime. Others counter it could occur when the sun grows only a few percent brighter, in just a few hundred million years. Both groups believe that when the steam first begins to fill the atmosphere, it will shield the Earth and its oceans from a portion of the sun's rays. But they can't agree on exactly how much sunlight will be reflected back by this emerging layer of vapor.

Scientists point out other, less predictable threats to life's future. For instance, every 24 to 30 million years the Earth may be bombarded by a shower of comets; roughly every 50 million years the Earth collides with an asteroid more than six miles in diameter—something a little smaller than Washington, D.C. Either type of collision could throw millions of tons of dust into the atmosphere and plunge the surface of the Earth into pitch-darkness for up to several months, wreaking havoc on the food chain. Researchers now widely believe that about 65 million years ago a comet shower or an asteroid wiped out the dinosaurs, along with over half the species living on Earth. (The effects of such an impact appear to be strikingly similar to what many scientists expect will occur in the aftermath of a full-scale nuclear war.)

There are other objects in outer space that could cause problems without crashing into Earth. Astronomers calculate that every 50 million years or so, a dying star more than 10 times as massive as the sun explodes close enough to the solar system to zap the Earth with possibly enough cosmic rays to kill off humans and other animals. Theoretically, a more distant supernova could still be powerful enough to engulf the Earth with nitric oxide, created when particles from the stellar explosion hit Earth's atmosphere. This nitric oxide could destroy the ozone layer, which shields Earth from the sun's otherwise deadly ultraviolet rays, and cause mass extinctions.

Lest this all seem too gloomy, remember that, in a sense, life has always bounced back after past catastrophes. Species may come and go, but living things have existed continuously almost since the creation of the solar system. They have endured a dim young sun, previous comet and asteroid impacts, and nearby supernovas, and still we're here to speculate about the future.

—*Steve Olson*

Part VI
THE UNIVERSE

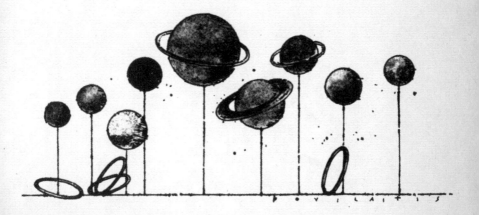

The Young Meteorites: Origin Unknown

Day after day, the Earth gets showered with tons of meteoritic debris. After analyzing about a thousand chunks that survived a blazing journey through the atmosphere, scientists have concluded that most meteorites are remnants of the solar system's birth, pieces of rock or scraps of iron and nickel that are about 4.5 billion years old.

But eight of these meteorites, collected from locations around the world, defy that description. Grouped according to composition, these shergottites, nakhlites, and chassignites—or SNC (pronounced snick) meteorites—share a common historical feature: All came from molten rock, or lava, that formed as recently as 1.3 billion years ago. What member of the solar system could possibly be the parent of these peculiar newcomers?

Most meteorites are thought to come from the asteroid belt, which lies between Mars and Jupiter. But the asteroids had long cooled from the solar system's fiery beginnings by the time the grayish SNC rock was born. Planetologists quickly nixed the moon for the same reason.

Deducing how these space travelers might have departed their birthplace has made it possible to rule out other sources. Even if a meteorite could form on the giant outer planets, Saturn and Jupiter, it would never reach an orbit that the Earth crosses in its journey around the sun. Venus is much closer, but its gaseous atmosphere is so dense that escape seems impossible.

In the late 1970s, several meteorite experts went out on a limb to nominate Mars as a candidate. They reasoned that the red planet's giant volcanoes were spewing lava about the same time the meteoritic rock was born. As the scenario goes, chunks of Mars' billion-year-old rock were blasted into space after a giant comet or meteor crashed into the Martian surface. Certain glassy fragments in one of the shergottites, in fact, indicate that the meteoritic rock was subjected to some kind of immense shock about 180 million years ago. Could that be the signature of an impact on Mars?

This idea was highly speculative at the time, interesting cocktail chatter at planetary science conferences. But the theory got a real boost when studies were conducted on a 17½-pound shergottite found in Antarctica in 1979. Donald Bogard of NASA's Johnson Space Center in Houston reported that the proportions of neon, argon, krypton, and xenon trapped within the glassy fragments of the Antarctic specimen turned out to be strikingly similar to the abundances detected on Mars by the Viking landers. Shortly thereafter Richard Becker and Robert Pepin of the University of Minnesota found that the trapped gases were greatly enriched with a rare isotope of nitrogen that also is characteristic of Mars. The researchers surmise that these gases were trapped inside the rock during the impact that shot the meteorite off Mars.

There are problems, however, even with Mars. Cratering experts have tested various impact models, and when they accelerate large rocks to Mars' escape velocity of three miles per second, the blocks of solidified lava either melt, vaporize, or get ground to a powder. More complicating, some of the young meteorites do not even appear to be shocked. Can something be blasted off a planet and remain unscathed?

Ann Vickery and Jay Melosh at the University of Arizona in Tucson have been examining various scenarios for how the meteorites might have been launched from Mars. They propose that the SNCs were ejected in a single event about 200 million years ago from a single as yet unidentified crater on the red planet. This idea has won support, but it remains impossible to verify without Martian rock samples. For the time being, the origin of the young meteorites remains unknown.

—*Marcia Bartusiak*

Where Did The Moon Come From?

That old devil moon continues to bedevil. "It's far easier to explain why the moon shouldn't be there," says M.I.T. geophysicist Nafi Toksoz, "than to explain its existence."

That may sound strange when the data amassed by the manned *Apollo* lunar missions should have settled, it seems, the age-old question of the moon's origin once and for all. But that just didn't happen. Even after years of intensive study, lunar scientists are still trying to recreate the story of how the moon came to be.

In pre-*Apollo* days, before the first human footfall on the lunar surface, three rival theories held sway: that the moon split off or fissioned from a fast-whirling Earth; that it was once a vagabond of space that had strayed too close and become ensnared in the Earth's gravitational field; that it was formed simultaneously with the Earth from the same primordial gas and dust that gave rise to the rest of the solar system.

But scientists at the 1984 Conference on the Origin of the Moon turned their backs on the three classic theories in favor of a new hypothesis—that the moon formed out of the impact of a huge object grazing the Earth.

The three previously contending theories partially account for some of what scientists know about the origin of the moon, but none can explain enough. In terms of the condensation hypothesis, if the moon and Earth formed at the same time in the same space "neighborhood"—and out of the same embryonic material—how did they turn out so different chemically? The chemical disparities include not only the low concentration of iron in the moon but also the absence of water and other volatile substances from the lunar surface.

As if the iron dilemma weren't troublesome enough, there are other facts that play havoc with the theories. The fission hypothesis says that the moon built bit by small bit from pieces or planetesimals, spun off or burned off from a hot, furiously spinning Earth. Few geophysicists, however, now believe in a fast-spinning infant planet. But even if that were the case, there's another hitch: The moon's orbit is inclined to be the plane of the Earth's spin, a circumstance inconsistent with a spin-off origin.

Physicists investigating the dynamics of close encounters of the planetary kind also question the basis of a capture theory. If nothing else, capture is a remarkably unlikely event. But a planetesimal passing close to the newly formed Earth might crash into it, which is what William Hartmann and Donald Davis of the Planetary Science Institute in Tucson first proposed in 1975. According to this scenario, the moon originated from what would be the most catastrophic event in Earth history.

So far this new hypothesis has withstood close scrutiny, and it continues to gain in popularity. But the riddle of lunar ancestry still remains unsolved, which brings to mind some lines by the 17th-century poet and essayist Joseph Addison:

Soon as the evening shades prevail
The Moon takes up the wondrous tale
And nightly to the listening earth,
Repeats the story of her birth.

Only there still remains the matter of learning how to decode all of her devilish whispers.

—Ben Patrusky

Why Is The Moon Ringing?

At McDonald Observatory, near Fort Davis, Texas, techni-
cians have been firing laser beams at the moon, using a
telescope with a 107-inch lens as a ray gun. They bounce
the beams off reflector panels that were left on the moon
by the astronauts. A bell rings in the observatory whenever
they score a hit, and an atomic stopwatch measures the
moon's distance at that moment to within four inches.

By analyzing measurements from many different nights,
the French astronomer Odile Calame and her collaborator,
J. Derral Mulholland of the University of Texas, have detected
echoes of an event that happened before the last Crusade.
Or at least they think they have.

Shortly after sunset on June 18, 1178, near Canterbury,
England, a group of men stood admiring the new moon, a
bright sliver hanging low in the west. As they described it
afterwards to the monk who recorded it for posterity, the
moon's upper horn suddenly split in two. A "flaming torch"
sprang out of the division, spewing "fire, hot coals, and
sparks." The moon writhed and throbbed "like a wounded
snake" and finally "took on a blackish appearance."

Reading the details eight centuries later, Jack Hartung, a

planetary scientist at the State University of New York at Stony Brook, thought that the men might have seen an asteroid smack into the moon. He found enough clues to predict an impact site: a spot on the back side of the moon, just beyond the northwest horizon.

Hartung went to the Lunar and Planetary Institute in Houston, Texas, and pulled out Russian and American photographs of the moon's back side. There, in the right place, he found a remarkably fresh crater, 12 miles across and twice as deep as the Grand Canyon, radiating white splatter marks for hundreds of miles. To gouge a hole like that, a boulder as big as the Houston Astrodome would have had to slam into the moon at a speed of 40,000 miles per hour.

An article about Hartung's work appeared in the *New York Times,* where it was spotted by Mulholland. Reasoning that such an impact would have set the moon to ringing like a gong, Mulholland and Calame checked their laser data to see whether the surface was still vibrating. It was. The ground moves back and forth with an amplitude of 80 feet.

"There was only one problem," says Mulholland. "We couldn't peg the date at 1178. We could say only that it happened sometime in the last few tens of thousands of years."

If the men of Canterbury did witness the collision, they have given us a valuable piece of information. The collision date, combined with the size of the crater and the shape of the present vibrations, should reveal whether the moon is squishy inside, just as thumping a watermelon gives a clue to its ripeness. And if we knew more about the moon's interior, we might be able, at last, to explain its origin.

Hartung and others have been devising ways of pinning down the date. An opportunity arose in 1976, when a Soviet spacecraft happened to land on one of the splatter marks and took a coring. But the top layers, unfortunately, were too jumbled for accurate dating. "Another possibility," says

Hartung, "is that the collision might have sprinkled moon dust on the Earth. Maybe it will show up in polar ice cores." Mulholland, meanwhile, thinks somebody ought to search the medieval literature of Ireland and western Africa, where the moon should have been visible that evening.

"As you can imagine, solving this mystery isn't one of NASA's highest priorities," says Mulholland. "But it's strange to think that if we really want to, we very well might be able. The past is never truly gone. It always leaves marks that you can see if you only look closely enough."

—Terry Dunkle

Why Do Planets Have Rings?

Of the nine planets known in the solar system, at least three and perhaps four are encircled by rings, all quite different. Saturn's three main rings, subdivided into thousands of ringlets, are made of ice, billions of space-going snowflakes, hailstones, and icebergs, each one pursuing its own separate orbit around the mother planet. The ring around Jupiter, discovered by the *Voyager 1* spacecraft in 1979, is a thin, tenuous, almost invisible band of smokelike dust. The nine rings of Uranus, discovered in 1977 when they eclipsed the light of a star, are very narrow, slightly off-center to the planet, and blacker than tar. No one knows what they are made of. Neptune's ring, announced in 1983 on the basis of old (and controversial) stellar eclipse data, may not even be real.

There is nothing surprising about the shape of the rings. Put a cloud of random junk in orbit around a planet, and the individual pieces will start to bang into one another. As they collide, the particles lose energy, and the cloud very quickly spreads out into a flat disk—in other words, a ring.

Disk or ring formation is a very common phenomenon in astronomy. The planets orbit in a single ecliptic plane today because they formed long ago in a flat disk of gas and dust around the newborn sun. The sun itself orbits in a disk of gas and stars known as the Milky Way galaxy.

The real mystery of the rings is the origin of the orbital junk. One possibility is that the stuff was primordial, that it formed at the same time as the planet itself. In the case of Saturn, which lies nearly a billion miles from the sun, the region was cold enough for ice rings to form when the planet was born.

Perhaps it happened something like this: As Saturn condensed out of the gas of the solar disk, it captured its own smaller disk. As the disk cooled, crystals of ice and lumps of rock formed. Those fragments on the outside collided and coalesced into the icy satellites we see today. But those fragments on the inside were within the so-called Roche limit, where the gravitational forces of Saturn were just too strong. Every time they tried to coalesce they were disrupted. Eventually, they spread out to form broad, translucent rings. (The rings do, in fact, straddle the Roche limit.)

An alternative and much older theory says that the rings formed from the pieces of a small, icy satellite that wandered inside the Roche limit and broke apart from the gravitational stress. Another version says they formed from a satellite demolished by the impact of a comet.

Satellite breakup may also explain the rings of Uranus, where there is some indirect evidence for moonlets orbiting among the rings.

Jupiter's ring, wherever it comes from, has to be replenished regularly. The dust particles of the ring are so small (millionths of an inch) that radiation effects and the planet's magnetic field cause them to spiral into the planet very quickly. The two tiny Jovian moons, discovered by *Voyager 1* right at the ring's inner and outer edges, are logical candidates for that dust, but it is not obvious why the moons should be emitting dust. Some researchers think Io, Jupiter's volcanically active moon, is a more likely candidate for dust particles.

The problem with all these theories is that, even if they are true, they do not really explain the why of the rings. If Saturn has rings made of primordial ice, why doesn't Uranus? Why should the Uranian rings be so dark? Why are the giant gas planets the only ones to have rings? Why aren't there rings around the rocky planets of the inner solar system—Mercury, Venus, Mars, and for that matter, Earth? Did Earth ever have a ring?

The answer is—nobody knows.

—*M. Mitchell Waldrop*

Why Is The Red Planet Red?

The fourth planet from our sun was named Mars, after the Roman god of war, because early observers thought the planet suggested a warrior covered with blood or with rusty iron armor. Scientists of the Space Age scoffed at such primitive personification. When it comes to the mystery of Mars' color, however, neither myth nor machine has an unassailable answer.

In the early 1900s, some astronomers thought Mars appeared redder at some times than at others because a springlike melting of its polar ice caps provided moisture for a surge in growth of Mars' reddish vegetation.

But by 1971, *Mariner 9* had orbited Mars photographing a dry, dusty, rock plain pocked with meteorite craters. Scientists soon turned their attention to the soil of Mars. Was it a pink, feldsparlike mineral? Was it a compound unknown on Earth? Such a compound was postulated, and dubbed "carbon suboxide." It was described as having plastic-like qualities.

In 1976 two Viking landers were sent up equipped with instruments to analyze the alien soil and send the information to Earth. But instead of settling the question of Mars' color, the landers have deepened the mystery, hinting that the last laugh may go to those ancients who thought it was iron or blood.

"The most believable theories now say that some sort of iron oxidation, or rusting, has gone on," says planetary scientist Ben Clark, member of a team that has studied Viking's analyses of Martian soil. "Martian soil, which is magnetic and has the consistency of dust, could be magnetite, an iron oxide mineral that is usually black," says Clark. "Mars' atmosphere has the capability to further oxidize this mineral, changing it to the red state."

But the theory of a rusty Mars presents some unanswered questions. Why, for instance, is Martian soil magnetic when most red, iron-rich minerals on Earth are not? "The dust grains could be black inside and red only on the outside, where they have oxidized," says Clark. But he is quick to admit that magnetite is only one of numerous possibilities. "It could be red salt," he says.

Perhaps, goes another theory, Martian dust is red throughout. In an eerie avatar of the ancient belief that Mars was blood-covered, two Princeton University geologists, Robert Hargraves and Bruce Moskowitz, think that Mars did indeed "bleed" from meteor bombardment of its surface, which they believe is composed of nontronite, another iron-rich mineral.

"The Viking data indicate the planet's surface may be nontronite," says Hargraves. "But on Earth, nontronite is neither magnetic nor red. It is usually a yellow-green mineral formed by hot springs boiling up from under the ocean floor." So how did Martian nontronite become red and magnetic?

Hargraves and Moskowitz theorize that eons ago, meteors of all sizes plowed through Mars' thin carbon dioxide atmosphere and buried themselves in the planet's surface. The tremendous shocks produced enough heat to transform areas of the nontronite surface to a red magnetic phase. Simultaneously, the meteors pulverized those areas into fine dust, which was gradually dispersed over the planet by strong winds.

In arriving at their theory, the two geologists experimented with earthly nontronite to see if heating it would change its color and magnetize it. They found that five minutes of heat at 1,652 degrees Fahrenheit produced red, magnetic nontronite. "Our theory is that Martian volcanic lava interacted with the water that once existed on Mars to form a nontronite crust," Hargraves says. "Meteors and wind painted that crust red."

The Princeton theory of meteor bombardment may have some holes in it. "Nontronite formation requires a lot of water," says Clark. "We're not sure how much water was ever there. We just won't know why the planet is red until we get a sample of its soil."

Meanwhile, the red warrior lies entombed in icy temperatures that fall to −140 degrees Fahrenheit. The chemistry behind its color is a Martian secret, and Earthlings may never really know what covered its face with red, magnetic makeup.

—*Calvin Allen*

Part VII

MATHEMATICS AND PHYSICS

Are No Two Snowflakes Alike?

About 600 cubic miles of snow cover the Northern Hemisphere, and in every cubic foot of the fluffy stuff there are about 18 million snow crystals. Considering that much more snow covers the entire Earth and that snow has been falling for perhaps 2.5 billion years, it's hard to believe that every single snowflake ever created is different.

"That's a question I've worried about quite a lot," says Charles Knight, a physicist at the National Center for Atmospheric Research in Boulder, Colorado. "I'd answer by asking: How different is different? Certainly you could find two that are very similar."

Of course, if a snowflake that fell on a corner of Antarctica a million years or so ago was identical to one that fell in your backyard yesterday, who would know? Determining whether it's even possible requires a knowledge of how snow crystals grow. And that, says Knight, is still somewhat up in the air.

Crystals, which fall individually or clump together to form snowflakes, are tough to study. Electron microscopes, for example, must operate in a vacuum. Put a snow crystal in a vacuum and poof! it's gone. And working at a temperature where snow feels comfortable usually means the scientist isn't. "We're a bit stymied," says Knight. "Almost nobody is working on this subject because of these problems."

But some of the basics have been worked out. Snow crystals are generally thought of as the intricately sculpted six-sided stars that adorn department store window displays and Christmas trees. But snow also comes in six-sided plates, hexagonal columns, needles, columns capped with stars, plate-capped columns that look like old-fashioned shirt studs, and a hodge-podge of irregular shapes. It all depends on the temperature of the cloud in which it's made.

Clouds are formed when a mass of warm air rises and cools, losing its ability to hold the water evaporated in it. The vapor condenses out of the air to form water droplets, which cool further into ice or remain as "supercooled" water. Other times the vapor condenses directly into ice crystals. If a cloud contains both ice crystals and drops of this "supercooled" water, a curious thing happens. The water droplet evaporates, and its vapor condenses onto the ice crystal. "It's like putting a cold can of beer next to a boiling teakettle," says Knight. "The vapor hits the can and condenses." In a cloud the vapor condenses directly into ice, and as more and more droplets evaporate and freeze, a snow crystal is born.

A water molecule's structure and electrical charge usually make ice crystals freeze into the shape of hexagons. Most snow crystals grow along the planes of this six-sided symmetry, and the general shapes they take depend on the temperature. At temperatures between 32 and 27 degrees Fahrenheit, plates form; stars begin to form at 10 degrees, and at three degrees plates form again. A snow crystal's intricate detail arises from minute variations in temperature and humidity that it encounters whirling about a cloud during its 15-minute birth.

Scientists have grown snow crystals in laboratories and catalogued the various shapes they take at different temperatures. But while scientists know which type of snow crystal forms at a certain temperature, no one has been able to explain *why* the crystal takes the shape it does. "The structure of an ice crystal is well known," says Knight. "But the mystery is how the molecules are packed together at the surface, which is where crystal growth occurs."

Which leads to the question at hand. Is it possible for two snow crystals to be identical? Considering that more snow crystals have fallen on the Earth than there are stars in the galaxy, it might seem surprising that a particular pattern would not be repeated. But each crystal contains trillions and trillions of water molecules that can be arranged in many different ways, and so it is perhaps more surprising that anyone would think two snow crystals could be the same. "It's like asking why no two people look exactly alike," says Richard Sommerfeld, a geologist with the U.S. Forest Service. "The real question is: Why should they?"

—*William F. Allman*

What's Going On Inside The Sun?

Some 400,000 miles deep in the sun's fiery core, at 27,000,000 degrees Fahrenheit, hydrogen changes into helium. Theoretical models of how the sun works predict that the by-products of this intense process are electrons, photons, and neutrinos. One of these, the neutrino, has a special beauty. Neutrinos alone escape the sun's core directly. Flying into space in prodigious amounts, these chargeless subatomic particles travel at about the speed of light to arrive on Earth eight minutes later. Neutrinos provide the only direct information physicists have about the inside of our sun—and about the future of sunshine.

But neutrinos are a bedevilment for physicists. In 1968 Raymond Davis of Brookhaven National Laboratory devised an experiment to detect solar neutrinos, but he counted less than a third of those that theory predicted. Based on the sun's temperature and other calculations, John Bahcall, a theorist at the Institute for Advanced Study in Princeton, New Jersey, estimates that about seven solar neutrino units should be detectable from the sun. Davis counted only two.

179

He refined his experiment; others checked it. But his numbers stand. "People have been comparing Davis' experiments and my theory for years," says Bahcall. "And no one's yet found a plausible mechanism to explain the discrepancy."

One possible reason for the difference in neutrino counts is that Bahcall may have miscalculated the sun's temperature. Davis counts only neutrinos with high energy, those that change the chlorine atoms in cleaning fluid, his experimental medium, into detectable radioactive argon. These high-energy neutrinos come from a rare reaction so exquisitely sensitive to the sun's temperature that a slight difference in temperature would have huge consequences for the number of neutrinos. "If you ask me to guess," says Davis, "calculations of the temperature are off a little bit." Bahcall, however, isn't convinced. "I don't see any way to get my numbers down to his," he insists. "If it weren't for the chlorine experiment, the standard model for how the sun shines would have no evidence against it."

Now if Davis could use gallium instead of chlorine, he could count lower-energy neutrinos, which come from a reaction much less sensitive to the sun's temperature. A neutrino count from a gallium experiment might come closer to matching a prediction based on solar theory. But enough gallium would cost $6 to $15 million, and the U.S. Department of Energy so far has not agreed to underwrite the experiment.

The problem instead may be that physicists are wrong about the neutrino. They know of three kinds, or "flavors." One flavor might be able to change, or oscillate, into another, as other particles do. Davis' experiment detects only the flavor leaving the sun. But if, during the 93 million miles between the sun and Earth, some neutrinos change flavor, Davis would understandably miss them.

Both American and European scientists have tried for

years to find out if one type of neutrino changes into another, experimenting over distances of half a mile. So far, says Felix Boehm of the California Institute of Technology, "no oscillations." He suspects, however, that neutrinos oscillate over much larger distances than can be tested on Earth, perhaps over the distance between our planet and the sun. If a gallium experiment continued to find too few neutrinos, oscillation would seem a likely candidate to explain the missing particles. Russian physicists will begin a gallium experiment in the next few years. "Any way that experiment turns out," says Davis, "we'll learn something about the solar neutrino problem."

Oscillation, however, has a nagging complication. Physicists used to be convinced the neutrino was massless. But in order to oscillate neutrinos must have mass. In 1980, a Russian group measured a tiny mass, so small it might be due to an error in measuring. Nearly 20 experiments, planned or running, are trying to repeat the Russian results. But it could be awhile before anyone knows the outcome of these tests.

Neutrinos come in immense quantities, and if it turns out that they do have mass, then together they might outweigh the stars. And if so, cosmologists say, their gravitational force will eventually halt the expansion of the universe, reverse it, and pull the whole thing into a black hole to end all black holes. Learning what goes on inside the sun might reveal the fate of the universe.

—Ann Finkbeiner

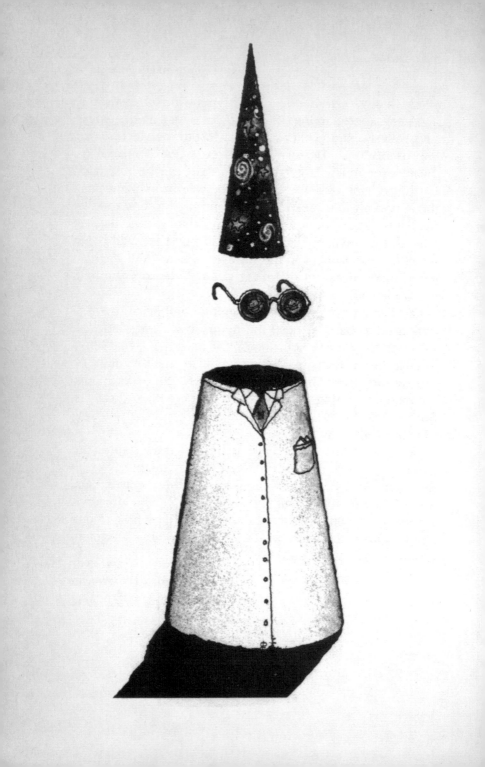

The Case Of The Missing Mass

The night sky is filled with stars and galaxies by the hundreds of billions. Yet they represent only a fraction of the matter in the universe. Much of the material in the cosmos appears to be invisible, and vast portions may not be made of matter as we commonly know it. Finding out what it is could, among other things, provide a clue to the fate of the universe.

Because this matter is invisible, it must be studied indirectly—by looking at the galaxies in a cluster, for example. Assuming the cluster is stable, its overall gravitational field must be strong enough to keep the fastest galaxies from flying away. Thus, by measuring the velocities of these galaxies and applying simple Newtonian physics, astronomers can estimate the total amount of matter in the cluster. When they do, they find that the visible galaxies in large clusters contain only a small percentage of what is calculated to be the total mass. Most of it is missing.

In individual spiral galaxies, meanwhile, each star orbits around the galactic center and is held in its path by the gravitational attraction of the material closer in. Over the last decade, Vera Rubin and her colleagues at the Carnegie

Institution of Washington have surveyed some 70 spiral galaxies and have found that the outskirts of each disk spin far too fast to be held together by only the stars that are visible. In other words, the outer reaches of spiral galaxies are held in place by "missing" matter.

Some astronomers have suggested that the elusive matter may be nothing but dark gas, dust, rocks, or even cold, dead planets: ordinary, prosaic matter that for some reason never condensed to form stars. Perhaps it is made of stars that are too dim to see. Or perhaps it contains something more exotic, such as tiny black holes—bits of matter so dense their gravity swallows up everything in reach, including light.

One intriguing possibility is that the missing matter is composed of massive neutrinos. First observed 35 years ago, neutrinos are electrically neutral, subatomic particles produced in certain types of radioactive decay. They were also formed in the Big Bang: in fact, neutrinos left over from that era are thought to outnumber the protons, neutrons, and electrons of ordinary matter by about 10 billion to one. On the average, every cubic centimeter in the universe contains some 450 of them. If no one feels particularly crowded, it's only because neutrinos display a ghostly indifference to ordinary matter: They can pass through the entire Earth as if it weren't there.

Until recently it was also thought that neutrinos were massless, like photons of light. But some of the new, unified theories of elementary particles suggest otherwise. And if neutrinos do have mass, then the cosmological consequences are profound.

If a neutrino were only one-thousandth of one percent as massive as an electron, for example, it would be heavy enough to congregate with others in the big clusters. Each neutrino would pursue its slow orbit among the galaxies, contributing infinitesimally to the invisible halo that keeps the cluster bound. Similarly, if neutrinos were somewhat heavier, they would be captured by individual galaxies and contribute to the massive halos there.

Finally, depending on their mass, there are enough existing neutrinos to halt the expansion of the universe. Their contribution of the cumulative gravitational pull of all matter could ultimately bring the galaxies back together in a crunch to match the original Big Bang.

And if it turns out that neutrinos don't have mass? No problem for the physicists: their unified theories have room for a whole beastiary of hypothetical objects: axions, photinos, "shadow matter," and what have you. Each of them would be a weakly interacting, massive particle—a WIMP— like the massive neutrino. And any one of them could explain the missing mass.

Unless, of course, none of the WIMPs really exist. But even then, the mystery won't go away. Something is out there.

—M. Mitchell Waldrop

In The Shadow Of Fermat's Last Theorem

In the margin of a book discovered shortly after his death in 1665, the brilliant French mathematician Pierre de Fermat made a claim that has vexed mathematicians ever since. Fermat scrawled an elementary theorem, and beneath it he wrote: "I have discovered a truly marvelous demonstration of this proposition that this margin is too small to contain."

If Fermat really did have a proof of this theorem, it died with him. An exhaustive search of his work uncovered no sign of a proof; and the concerted efforts of some of the world's greatest mathematicians have failed to produce one. Today, what is known as Fermat's Last Theorem remains the most famous unsolved problem in all mathematics.

The problem is based on the Pythagorean theorem familiar to all algebra students. The Pythagorean theorem, which unlike Fermat's Last Theorem is easily proved, states that when the lengths of the two shorter sides of a right triangle are squared and added together, the sum will equal the square of the length of the triangle's longer side ($x^2 + y^2 = z^2$). If x, y, and z are positive whole numbers, they make up a set known as a Pythagorean triple. For instance, the smallest Pythagorean triple is three, four, and five: three squared plus four squared equals five squared. To take a larger example, 3,398,640 squared plus 4,805,119 squared equals 5,885,569 squared.

In his famous marginal note, Fermat wondered whether other triples could be found if the numbers were cubed rather than squared, or if they were raised to the fourth power, or if they were raised to any power. He concluded that they could not. According to Fermat's Last Theorem, the only power, or exponent, for which such triples exist is 2. (In mathematical terms, $x^n + y^n = z^n$ has no solution in positive whole numbers, except when $n = 2$.)

Proving Fermat's Last Theorem means showing it to be true for *any* exponent. But in the face of profound difficulties, the mathematicians who followed Fermat fell back on proving the theorem for specific exponents. The first proof for the exponent 3 appeared about a hundred years after Fermat's death—the work of the great Swiss mathematician Leonhard Euler. Using mathematical entities known as imaginary numbers, Euler showed through a clever algebraic manipulation that it was impossible for the sum of the cubes of any two numbers to equal the cube of a third number. But Euler's proof contained a subtle, though in the end correctable, error. It was a sign of the problems to come.

By the 1840s mathematicians had proved Fermat's Last Theorem for every exponent up to seven. There progress faltered. The equations used in the proofs became so long and cumbersome they could be pushed no farther.

Then, in 1847, the German number theorist Ernst Kummer scored a major breakthrough. Using a new approach—a set of numbers known as the cyclotomic integers—Kummer managed to prove Fermat's Last Theorem for every exponent up to 100. In fact, at one point Kummer seems to have believed that he had produced a proof of the theorem for *any* exponent. But before he could publish his proof, he detected an unresolvable error in its logic.

Today, some mathematicians are trying to subdue the theorem with brute force, using supercomputers to calculate proofs for ever larger exponents. At last count, the theorem was proven for every exponent up to 125,100. If an exception to the theorem does exist, thus disproving the theorem, it involves numbers so large that no conceivable computer could calculate them. In a sense, then, Fermat's Last Theorem is empirically true.

But that doesn't satisfy the pure mathematician. To prove the theorem conclusively, it must be proved for each of an infinite number of exponents.

In 1983 a young West German mathematician named Gerd Faltings took an important step in that direction. In a 40-page proof that others have termed masterful, he showed that for each exponent, there could be only a finite—as opposed to infinite—number of exceptions.

Based on Faltings' program and the work of a Russian mathematician named A. N. Parshin, Yoichi Miyaoka developed a proof while working at the Max Planck Institute for Mathematics in Bonn, West Germany. Many mathematicians thought that Miyaoka's proof might be the long-sought solution to the 350-year-old problem. But after closer scrutiny, it appears unlikely that the Japanese mathematician has solved the Holy Grail of mathematics.

Given all this work, many mathematicians doubt that Fermat really did have a proof. They suspect that he jotted down his assertion based on a mistaken hunch, and that when he discovered his error he never went back to change his marginal note. But no mathematician can *prove* that Fermat was mistaken. In life as in mathematics, such proofs can be elusive.

—*Steve Olson*

Part VIII
NATURAL PHENOMENA

A Lake In Antarctica?

The new images of Earth from the *NIMBUS 5* satellite delighted NASA's scientists. For the first time they could photograph the planet without having clouds or the long polar night run interference between camera and continent. The images came in without a hitch. Europe looked beautiful. Asia was crystal clear. Then they took a good look at Antarctica.

Smack in the middle of the continent's frozen-in-winter Weddell Sea lay an unfrozen 186,000-square-mile lake, an area greater than California where no lake should have been. "Polynya," they murmured, a Russian word meaning "unfrozen water surrounded by ice," and then looked at the photographs again.

It seemed impossible. Polynyas weren't supposed to be that big. The ones that scientists had often spotted in the Arctic Ocean were tiny by comparison, and these were always made by the wind. The explanation was simple: If the wind were high enough to push the ice away at a rate of more than a few yards per second, then a polynya a few miles wide would form. But no wind on Earth was powerful enough to keep an area as large as that of the Weddell polynya clear of ice. Scientists decided that something else had to be at work as they learned more about the large polynya, which persisted for the entire winter and for three consecutive winters.

What could melt a hole larger than California through the Antarctic ocean's winter ice? One hypothesis was that currents moving over a rise on the ocean floor were causing a circular eddy to form. In the Antarctic, the water about 500 meters down is slightly warmer than the water at the ocean surface, because it is shielded from contact with the colder air. If the eddy created by the ocean floorscape were strong enough, its vertical motion would bring the warmer water up to the surface, where it would prevent the winter ice from forming.

A second hypothesis by Arnold Gordon, of Columbia University's Lamont-Doherty Geological Observatory, suggested that the answer instead came from above. In winter, when ice forms on the surface, it dumps most of its salt into the water beneath. Given the right circumstances, the under-ice water becomes denser and heavier than the warmer deep water, and the two types of water trade places. According to Gordon, this forms a deep but relatively narrow convective chimney about 30 miles wide; in it the denser top water sinks while the lighter and warmer deep water rises, keeping the ice from forming. But since the polynya covered 186,000 square miles one chimney alone couldn't do the job. There would have to be a lot of them, all going at the same time like a field filled with industrial smokestacks.

While Arnold Gordon's chimney-field convection process does seem plausible, the way the process stops is still anybody's guess. Contrary to what the satellite photos of real polynyas have shown, Gordon's best theoretical models have indicated that a convection field that managed to get itself started would then persist, that is, go on convecting forever.

An expedition of Soviet and American scientists found out otherwise when they set out to penetrate deep into the Weddell pack ice to try to learn more about the polynya firsthand. Thousands of temperature readings were taken. Cooperation between the Russian and American scientists was excellent. The only thing that failed to cooperate was the polynya, which didn't form that year.

Gordon and a colleague from NASA's Goddard Space Flight Center, J.C. Comiso, have been studying two smaller polynyas in an attempt to understand the Weddell polynya. They believe that melting sea ice is probably responsible for terminating the convection process of polynyas. The sporadic nature of the Cosmonaut and Maud Rise polynyas suggests that the size of a polynya is related to persistence—large polynyas are better protected from invading sea ice, which damps out smaller ones. But no one has an explanation yet for the mysterious and intriguing Weddell polynya.

—Dave Fleischer

The Movable Earth Puzzle

Bill Utterback was the first geologist on the scene at the isolated ranch in north-central Washington State shortly after hearing about the keyhole-shaped chunk of earth. It measured seven by 10 feet, weighed more than one and a half tons, and lay 75 feet from a hole of exactly the same dimensions. There was no doubt that the turf came from the hole—a hole that hadn't been there a month earlier, when ranch hands had gone to round up cattle grazing on the remote windswept plateau—but no one had any idea what had moved it.

Utterback, a mining consultant for the Colville Confederated [Indian] Tribes, chuckles a little nervously when he talks about what he saw that day. "The spot is pretty spooky to begin with," he says, referring to the area known as Haystack Rocks for the house-sized boulders deposited there eons ago by retreating glaciers. "It's bleak and cold and it gets dark real fast up there." He found no bulldozer tracks in the soil, no scorched earth to suggest an explosion. The grass roots of the plug had been ripped out of the ground, not cut. And a trail of dirt lay in a curved path from the hole to the slab, which was rotated about 20 degrees counterclockwise from its original position.

The event is so peculiar it does not fit into any geological category, leaving scientists wondering even what verb to use to describe it. Some who had been to the site have said the plug "popped" or "floated." "People seem to like to call it 'the earth cookie,' " says Utterback, since the chunk has two-foot-high, nearly vertical sides and a flat bottom that made it look as if it had been cut out of the ground by some monstrous kitchen implement.

That scenario may turn out to be as likely as any. Underground methane eruptions looked like a possibility for a while, until it turned out that the gas is rare in the area. Lightning was considered, and even a freak tornado. An exploding meteor might have created an enormous updraft over the spot, but there were no signs of violent activity—which ruled out most of the other explanations. Because the area lies in a slight depression, one geologist, in desperation perhaps, suggested that on a cold night the chunk froze and, following a torrential downpour, rose like an ice cube to the surface of the lake that had formed over it. Then it simply drifted away. "Yeah, I've heard that one," says Utterback. "It's impossible. How could you even melt the ice and get rid of that much water in a couple of weeks?" Besides, he says, almost no rain fell on the area during that period.

Some scientists have attempted to link the event to a nearby earthquake that occurred nine days before the puzzle piece was found. At the epicenter, 20 miles distant, the quake measured 3.0 on the Richter scale. Indeed, rare reports of objects raised out of the ground during or after earthquakes date back to 1797, when vertical shock waves supposedly hurled Ecuadorian citizens 100 feet in the air. A century later in Assam, India, tremors tossed boulders up, but they came straight back down. As recently as 1978, a 3.5-force disturbance in Utah was blamed for creating a depression two feet in diameter by throwing fist-sized clods of earth as far as 14 feet away. But a rain of clods doesn't compare to a one-and-a-half-ton colossus.

Still, for want of a better explanation, U.S. Department of Interior geologist Gregory Behrens slightly favors the earthquake theory. Just below the topsoil at the site lies a hard layer of bedrock that curves downward slightly to form the shape of a shallow bowl. "If there's good, hard wave transmission through the rock," says Behrens, "it could focus in the center much the same way that if you rap hard enough on the outside of a bowl of water, you'll see a disturbance in the center."

But geophysicist Stephen Malone of the University of Washington says, "I just can't see any way that that could happen. The ground motion could not have been strong enough from a magnitude-three earthquake. Waves dissipate quickly and this one was too far away, and there wasn't enough energy to begin with."

There are a growing number of other theories, from clandestine Defense Department operations to tiny "black holettes" colliding with the planet, but none yet that can fully explain the voyage of the restless turf.

"What's really got me scratching my head," says Utterback, "is not so much that it cleared the ground, but the amount of energy it would need to resist gravity over such a distance. If it's a hoax of some kind, why do a whole lot of work in the middle of nowhere? Why do it at all?"

—*Patrick Cooke*

What Made Those Mounds?

At the fenced edge of a natural preserve area 75 miles south of Seattle, nuthatches chirp busily, and the wind whispers through the bordering trees. Inside the 445-acre enclosure, large mounds of earth, 20 to 30 feet across and 30 to 50 feet apart, rise from the Mima (pronouncd MY-mah) Prairie. The origin of these mounds has remained a puzzle for more than 140 years, a mystery compounded by the fact that the formation, or microrelief, also exists or has existed in California's San Joaquin Valley, Colorado, Oregon, and eastern Oklahoma and Texas. The mounds are referred to by names as whimsical as hog wallows, gopher mounds, and pimpled prairies.

While mountains, prairies, and rolling hillsides are common, Mima-like mounds are not. The pimpled or Mima microrelief usually can be found in areas with a thin layer of topsoil overlaying coarse gravel, hardpan, or solid rock. The mounds range in height from one to eight feet and from 10 to 60 feet in diameter. These piles are a mixture of dark-colored clay, sand, and pebbles, and some have a high organic content of 21 to 30 percent. They're homogeneous, not layered, and contain no animal or human remains but often have traces of pollen, indicating the presence long ago of plants buried in a boggy environment.

How these mounds were formed is as mysterious today as when Sir James Douglas of the Hudson's Bay Company first saw them in 1840. Many early theories have been discarded; they are not volcanic vents, Indian burial sites, or ancient sea bottom formations made by whirlpools. Today speculation centers around gophers and glaciers.

Victor Scheffer, a biologist formerly with the U.S. Fish and Wildlife Service, has championed the cause of the gophers for more than 40 years. In his theory, a group of common pocket gophers once lived on the Mima Prairie. Because the area had only a thin coating of topsoil and the gophers couldn't make their burrows in the underlying coarse gravel, they tended to gather the soil into mounds and make their homes there. Ten gophers on one acre of land can move four to five tons of earth a year. So it's not surprising, Scheffer claims, that over a few centuries, or even decades, they raised up mounds taller than a man and wider than a living room.

Though gophers don't live on the Mima Prairie today, the area does lie within their range, and they could have moved enough topsoil to make them. Some even claim to have found traces of burrows in the mounds, now filled with dirt and gravel washed in by the ubiquitous Washington rains.

Just as plausible an explanation is the periglacial, or near glacial theory, originally proposed in 1952 by Reuben C. Newcomb, a geologist formerly with the U.S. Geological Survey. According to Newcomb and others, as the great glaciers retreated from western Washington 15,000 years ago, the exposed land alternately froze and thawed through the years, creating a network of ice wedges in the earth. As the wedges swelled, the dirt trapped in between was compressed into polygons, which formed mounds when the ice melted. Meltwater runoff from nearby glaciers further eroded the dirt between the mounds. later, some have suggested, the area became a bog; the still waters preserved the mounds from further erosion and weathering. This periglacial action occurs today in semifrozen areas of Alaska and Siberia.

Each theory explains mound microrelief in some locations but not in all. But the explanations are just educted guesses. Washington's Mima mounds could have been formed by glaciers or by gophers. Earth clustering around the roots of now vanished bushes might have formed the hog wallows of Visalia, California. There is evidence that the pimpled mounds in Pipkin Marsh, Jefferson County, Texas, were formed by processes of erosion and redeposition. More baffling, some of those mounds contain refuse of human habitation. And not far from the pimpled prairies of eastern Oklahoma are genuine Indian burial mounds.

The puzzle remains unsolved. The mounds still stand, mute witnesses to nature's ability to confound the mind of man.

—*Joel Davis*

What Was That Noise?

The Indians told of mountains that thundered, but Lewis and Clark scoffed. In their journal the explorers dismissed the story as "some superstition or perhaps a falsehood." Then, in June of 1805, as their expedition approached the Rocky Mountains in central Montana, they began to reconsider.

"We have repeatedly heard a strange noise coming from the mountains," they wrote. "It is heard at different periods of the day and night, sometimes when the air is perfectly still and without a cloud, and consists of one stroke only, or of five or six discharges in quick succession. It is loud and resembles precisely the sound of a six-pound piece of ordnance."

For at least 2,000 years, from the delta of the Ganges River to the shopping malls of Paramus, N.J., that same strange noise has been booming forth on land and sea. The sounds go by dozens of local names—*mistpoeffers* ("fog belches") off the coast of Belgium, Cornwall thumps in Ontario, Barisal guns in India, *brontidi* ("like thunder") in Italy, sea farts in the Bay of Fundy—but they are so similar that scientists refer to them all as brontides. They usually resemble distant thunder or artillery, occur on a clear day in a place where there isn't supposed to be any artillery, and are followed by the sound of humans unsuccessfully trying to explain what happened.

Not surprisingly, Aristotle was one of the first to propose a hypothesis. Noting that "in some places subterranean noises are heard unaccompanied by earthquakes," he guessed that winds were responsible. Subsequent theorists connected the sounds with collapsing riverbanks, falling meteors, ghosts, cracking ice, bears rolling boulders down mountainsides, "the bursting of rich mines of silver" (Lewis and Clark's explanation), and the "expressions of displeasure" of an Indian god who "was angry because the English god intruded upon him" (an American Indian explanation).

After a puzzling series of "airquakes" was heard in New Jersey and other places along the East Coast a decade ago, investigators found that most of them were distant aircraft supersonic booms transmitted by unusual atmospheric conditions. But some appeared to be natural booms, which renewed scientific speculation about brontides. Currently the two most popular theories involve earthquakes and explosions.

A contributor to the *Monthly Weather Review* in 1898 wrote that brontides "may result from earthquakes too slight to be otherwise perceived," and seismologists have recently proved him right. They find that audible booms can be produced by an earthquake's compressional shock wave, which causes the ground to vibrate up and down, sending sound waves into the air just like a loudspeaker. By using a seismometer, a microphone, and a tape recorder during a 1975 earthquake in California, David Hill of the U.S. Geological Survey recorded a boom from a shock wave that caused no obvious tremors. "The sound was a soft rumble, sort of like distant thunder," says Hill, who thinks that most reported brontides, including those heard near water, have been associated either with earthquakes or with distant explosions transmitted by freakish atmospheric conditions.

Thomas Gold isn't so sure. Gold, an astrophysicist at Cornell University, agrees that many brontides have been caused by earthquakes. But his calculations show that to produce a very loud boom, an earthquake must be strong enough to be felt. The earthquake theory, he says, therefore doesn't account for loud booms unaccompanied by earth tremors. Nor does it account for the many reports that mention booms along with flames. So Gold and an associate at Cornell, Steven Soter, have revived a theory that was used at the turn of the century to explain the "lake guns" heard in upstate New York. They suggest that some brontides are caused by high-pressure natural gas that escapes rapidly from the gound and explodes in the air, perhaps sometimes igniting into flames.

"That hypothesis has no merit," says Donald Stierman, a geophysicist at the University of California at Riverside who has studied the flow of gas underground. "Except in very special circumstances, like geysers or mud volcanoes, gas just doesn't escape from the ground quickly enough to produce booming sounds." He thinks that virtually all natural booms are caused by earthquakes, and he rejects the gas theory by arguing that "there is no need to invoke different physical processes for a single phenomenon."

To which Gold and Soter reply, "It is not clear how he knows that we are, in fact, dealing with a single phenomenon." Neither the arguments nor the noises seem likely to abate soon.

—John Tierney

What Are Those Lights?

By day, highway U.S. 90 between Alpine and neighboring Marfa is indistinguishable from the rest of west Texas. It is 26 miles of emptiness, peppered with mesquite and cholla. At night there are cars along this same route, but they are parked, their occupants staring off into Mitchell Flat to the south. These people are looking for the Marfa lights.

There, in the distance, a disk of light moves across the field. Suddenly it bounces, then seems to hover for a moment, or it zigzags around. Maybe it divides into two disks, each of which cavorts around. Sometimes there are several at a time. Other times they vanish entirely. To most witnesses, they seem far away. Others report being chased by them. The descriptions differ, but no one disputes that the lights are real. They can be seen almost any night.

No one knows what is different about that handful of miles around the Nopal, Antelope Springs, and Escondido ranches, but the lights aren't seen anywhere else. There is no want of theories about what they are. Most, however, are nonscientific. While few people believe in Indian ghosts anymore, lots of folks are credulous enough to associate the lights with supernatural forces or extraterrestrial visitors.

According to one tale, World War II aviators from the Marfa air base—right in the haunted area—went out at night and bombed the lights with sacks of flour to mark the spots for investigation. It is a common story but a false one. No trace of the markers was ever found. Fritz Kahl, who then was an instructor at the base, says flatly. "It never happened. Sure we chased them, but no one ever dropped flour sacks."

At least four possibilities do not require belief in the supernatural. Phosphorescent minerals, for example, absorb ultraviolet light. Even after a radiation source is gone, these minerals will glow for a time as energy is released from within. Geologists who have examined the Marfa area, however, declare that there are no phosphor deposits in the vicinity.

In some localities, methane (commonly known as swamp gas) escapes from the ground and produces eerie lights, probably from spontaneous combustion. The confirmed cases tend to be swampy areas where the methane is produced by rotting vegetation. There hasn't been a swamp in West Texas for several million years. And there is no known gas near Marfa.

Kahl and Curt Laughlin, the superintendent of nearby McDonald Observatory, have considered another source—the electrical phenomenon called St. Elmo's fire. But what is there about one isolated patch of rangeland that would generate high-voltage static electricity to the exclusion of the similar pasture for miles around?

A more attractive hypothesis has to do with the peculiar paths light can take. Astronomer Eric Silverberg speculates that the Marfa lights may be simply automobile headlights carried over great distances and along writhing paths by atmospheric tunneling, also known as the Novaya Zemlya effect. Large and abrupt temperature variations above the

Earth cause sharp changes in the density of the air, bending light in funny ways.

The headlight theory, however, encounters two problems. South of U.S. 90 and east of Marfa is empty country for a long, long way. In addition, the first recorded report of the lights was made by Robert Ellison—in 1883. But, Silverberg is quick to point out, starlight may also be subject to atmospheric tunneling. Stars appear a likelier source for the phenomenon at Marfa, but no one has proven it.

Out on the range, the lights continue to bounce, divide, and disappear. Every theory yet proposed seems to have flaws. Perhaps a definitive test would be to analyze the color spectrum of the lights, which would determine what chemical elements are involved. To make such measurements, however, would require a large portable telescope and a suitable spectrograph. No one has yet volunteered.

The residents accept the lights as a fact of life. Laughlin's response is perhaps typical of local feeling. "I hope we never find out what the Marfa lights are. Some things in our lives need to remain mysteries."

—*J. Derral Mulholland*

How is Lightning Sparked?

When Benjamin Franklin bravely flew his kite during a thunderstorm in 1752, he proved what most other scientists of his day only suspected: Lightning is indeed an electric spark. But today, scientists still don't know how those spectacular flashes get started.

Why does a billowy cloud suddenly become a high-voltage generator spewing out sparks that can stretch many miles across the skies? The central mystery is how huge amounts of positive and negative charges develop and polarize in portions of what is, at first, an electrically neutral cloud.

Finding the answer, however, has proved to be a lot harder than flying a kite. "A thundercloud is such an enormous thing, you really can't get your hands on it," says Earle Williams, a geo-physicist studying lightning at the Massachusetts Institute of Technology.

Many have tried. Scientists have taken death-defying balloon rides on the edge of raging thunderstorms to look for answers. They've flown airplanes around and through thunderclouds. Rockets have even been used to trigger lightning. But so far these efforts have provided only glimpses of what happens inside a thundercloud.

In most cases, a thundercloud must be more than two miles tall before it produces lightning. Its upper regions are usually bitterly cold; in general, the taller the cloud and colder its top, the fiercer the storm. A thundercloud is also characterized by intense electric fields and powerful updrafts and downdrafts. The negatively charged region generally lies in the lower section of the cloud, whereas the colder, upper region is positively charged. If this imbalance of electrical charges is great enough, it overcomes the air's insulating capacity and results in a lightning flash—the discharge of electricity between the cloud's two oppositely charged regions or between the negatively charged lower region and positive charges on the ground.

But what causes the separation of electrical charges? A thundercloud generally produces rainfall, and for this reason scientists have long suspected that precipitation is the driving force in its electrification. Under the most popular theory, precipitation in the form of large raindrops or ice particles falls downward carrying a negative charge. Lighter, positively charged particles, such as cloud droplets and ice crystals, accumulate in the upper part of the cloud. The resulting electric field may—or may not—produce a lightning flash.

MIT researcher Williams and Roger Lhermitte at the University of Miami have been using radar to determine if there is a change in the rain's velocity following a lightning bolt. Their premise is that if falling raindrops create the cloud's electrical energy, then they are moving against the atmosphere's electrical forces—a fact that should slow its descent to Earth. After lightning flashes, however, the electrical field is weakened and precipitation should fall more quickly. But in most thunderstorms, Williams and Lhermitte could detect no change in velocity.

John Latham at the University of Manchester, England, has a version of the falling precipitation theory suggesting that the primary charging process involves the collision of small hail with ice crystals or extremely cold water droplets. The falling hailstones splinter, and the positively charged bits of ice blown off them concentrate in the upper region of the cloud, while the heavier, negatively charged hail continues falling and creates the thundercloud's lower, negative pole.

But if falling precipitation provides the electrical energy, why does lightning frequently occur before—rather than during or after—a downpour? And why is there lightning in those rare clouds where the temperature is above freezing, or in dust storms or volcanic eruptions?

Bernard Vonnegut, at the State University of New York at Albany, theorizes that a cloud's charges actually originate outside the cloud. Excess positive charges in the atmosphere get sucked into the upper regions of the cloud, attracting negative charges from the upper atmosphere. The negative charges attach to cloud particles that are continually swept downward by drafts into the cloud's lower regions. The vigorous upward and downward air currents could be the mechanism for building up the acquired charges and keeping them separated.

But, as with the other theories, nothing is certain. Williams, Vonnegut, and Latham all concede they need to know far more about the inner workings of a thundercloud before they can explain lightning. And if they solve that mystery, there are still plenty of others.

No one knows, for example, why lightning assumes those devilish, crooked patterns. Or why it is more frequent over land than over water. Or why it usually hits the highest object in the area—but not always. Or. . . .

—*Mary Fiess*

What Is Ball Lightning?

During a thunderstorm in the English town of Smethwick, a housewife noticed a ball of violet light hovering over the kitchen stove. Rattling faintly, the light floated across the room and touched her. Instinctively she brushed it away. An explosion shook the room, and the ball vanished, burning a hole in her dress and scorching her underwear. The woman herself was unharmed.

It was a typical case of ball lightning. A luminous sphere, the size of a grapefruit, usually red to yellow but occasionally another color, wanders around a room or out in the street, obeying neither wind nor gravity, and after a few seconds, disappears, sometimes with a pop or a boom. Nobody is badly hurt.

Millions have witnessed ball lightning, and thousands have published what they saw. But scientists still have too little information to say what it is. A few think it is only a spot before the eyes, an afterimage from a nearby flash of lightning. Others say it's a cloud of vaporized metal from whatever was hit. Still others claim it's a volume of luminescent air, a swarm of glowing pollen grains, a bubble of burning methane, a dollop of anti-matter, a throng of electrified gnats.

James Dale Barry, a senior scientist at Hughes Aircraft in El Segundo, California, has evaluated each theory in the light of 1,000 eyewitness reports. He says ball lightning is probably luminescent air.

If that is true, ball lightning works something like this: A bolt of ordinary lightning pumps energy into electrons orbiting molecules of air. The electrons jump into higher energy states, linger awhile, and fall back, giving up the energy as tiny flashes of light. In man-made luminescence, the flashes can be triggered all at once, as in a laser, or in constant succession, as in a fluorescent lamp. Ball lightning behaves like the fluorescent lamp, with one exception: Somehow it continues glowing after the power is shut off.

But the theory doesn't explain ball lightning's shape. What keeps the ball spherical? Why doesn't it disperse like smoke?

The answer would interest people working on atomic fusion, the clean source of power that many believe will end our energy shortage. Before fusion can succeed, we must find a way to confine incredibly hot gas in midair, away from the walls of its container.

"I'm not saying research in ball lightning will solve all the world's problems," says Barry. "But I don't think we should underestimate it either. It was research in ordinary lightning that led us to the electric light, the telephone, the radio, and the computer."

Unfortunately, ball lightning theories rest largely on sparse and unreliable data. "The average witness is too startled to notice any but the roughest details," Barry says. "He remembers color and size, general direction of movement, a couple of other things, and that's it. Most of the reports tend to repeat the same things."

Scientists have tried other means of getting data. In France they have fired rockets that lift trails of wire into thunderclouds, enticing lightning to strike close enough to be watched for signs of globularity. In England they have calculated the energy in the ball that hit the lady in Smethwick from the melting point of polyester and the size of the hole in her dress. But real progress will have to wait until witnesses sharpen their eyes.

"Try to stay calm," Barry advises, "Ball lightning has never killed anyone as far as I know, although I wouldn't urge anybody to touch it. Watch the ball closely. Look for details. Can you see into the ball? Does it have an interior structure? Is it precisely the same color all over? Does it change in any way when it passes a metal object? If it touches anything, does it leave a residue you can collect?" And, Barry adds "Most important, can you get a good picture of the ball? Nobody ever has."

—*Terry Dunkle*

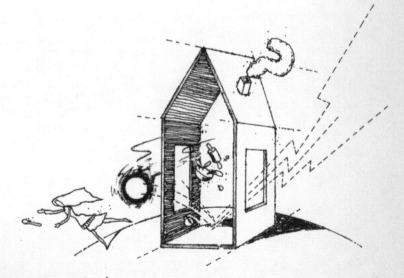

Index

The Day That Lightning Chased The Housewife

Designed by Gisèle Byrd

*Composed by Coghill Composition Company, Richmond, Virginia
in ITC Garamond
with display heads in Britannic Roman*